P9-ECT-632

THE BIRTH OF
ATHENIAN DEMOCRACY

The Birth of
Athenian
Democracy

The Assembly in the Fifth Century B.C.

CHESTER G. STARR

New York   Oxford
OXFORD UNIVERSITY PRESS
1990

Oxford University Press

Oxford   New York   Toronto
Delhi   Bombay   Calcutta   Madras   Karachi
Petaling Jaya   Singapore   Hong Kong   Tokyo
Nairobi   Dar es Salaam   Cape Town
Melbourne   Auckland

and associated companies in
Berlin   Ibadan

Copyright © 1990 by Oxford University Press, Inc.

Published by Oxford University Press, Inc.
200 Madison Avenue, New York, New York 10016

Oxford is a registered trademark of Oxford University Press

All rights reserved. No part of this publication may be reproduced,
stored in a retrieval system, or transmitted, in any form or by any means,
electronic, mechanical, photocopying, recording, or otherwise,
without the prior permission of Oxford University Press.

Library of Congress Cataloging-in-Publication Data
Starr, Chester G., 1914-
The birth of Athenian democracy :
the Assembly in the fifth century B.C.
Chester G. Starr.
p.   cm.   Includes bibliographical references.   ISBN 0-19-506586-7
1. Athens (Greece)   Ekklesia.
2. Athens (Greece)—Politics and government.
I. Title.   JC79.A8S7   1990   328.3′0938′5—dc20
90-31265   CIP

2 4 6 8 9 7 5 3 1

Printed in the United States of America
on acid-free paper

To
Peg and Ira Campbell

Preface

The present work considers an aspect of Athenian constitutional structure which has been too little treated. In dealing with its development and some of its intricacies I have been much assisted by the thoughtful suggestions of two anonymous critics; authors, I fear, do not always appreciate how much time and attention their colleagues may devote to such analysis.

Out of deference to that mythical character, the general reader, I have used the terms assembly and council rather than *ecclesia* and *boule*, but if there be any such who turn to these pages I trust that he or 'she will forgive the occasional appearance of Greek technical terms. Precision is sometimes necessary. On the other hand I have at certain points diverted the flow of the argument to include brief treatments of major turning points in the general history of the fifth century, including especially the Persian and Peloponnesian wars and the rise of the Athenian empire. Not everyone may feel deeply conversant with these events, but they had wide-ranging effects; institutions do not mature in a vacuum. But always my focus is on the development of the Athenian assembly as the vital element in the determination and supervision of state policy.

It remains to note that Ira Campbell and I have known each other since grade school over 60 years; not often do two people maintain such a lasting friendship.

Ann Arbor, Mich.   Chester G. Starr
February 1990

Contents

THE BIRTH OF
ATHENIAN DEMOCRACY

Introduction

Athenian democracy is a popular subject on which new books continue to appear regularly. None of these eulogies, oddly enough, devotes much attention to the issues as to how the assembly operated or developed its structure even though it has rightly been called the "dynamo" of the Athenian political system.[1] Studies abound on other parts of Athenian government such as the council of 500, the board of 10 generals, the secretaries, the treasurers, and other officials, but the most recent monograph on the *ecclesia,* as Mogens Hansen has pointed out, was published in 1819.[2] Hansen himself has written a detailed and judicious study of the assembly, but as a cautious scholar confines himself to the period 355–322, the age of Demosthenes and other orators who produced a reasonable amount of information.[3] This limitation unfortunately restricts the value of his analysis when one turns back to the fifth century, the period in which Athenian legislative procedures were in process of creation.

Another valuable work, by Martin Ostwald, does bear exclusively on the era in which we are here interested, but it is more concerned, like most studies, on Athenian democracy in an abstract manner, and the assembly appears in his pages only sporadically in its practical operations. Moreover, Ostwald's primary purpose is to prove that fundamentally sovereignty shifted from the whim of the people to governance by formal laws. This theory has

3

constitutional merits, but his sweeping conclusion that "Council and Assembly receded into the background in matters of internal policy, and the jury courts held center stage" accords ill with the evidence.[4] In the fourth century the assembly still determined foreign and domestic policy as its members saw fit.

In short, the questions as to when the assembly emerged and also the reasons and time when it had evolved to the point that it could play its role effectively remain almost untouched. Without such a survey our view of Athenian democracy must remain sadly incomplete, so an effort to penetrate carefully into the fifth century B.C. appears highly desirable. The evidence for this period is more limited than one might wish, but the main lines of change at least emerge as one considers the testimony of surviving decrees and a variety of literary materials, including historical and philosophical studies as well as the valuable poetry of Solon even if detailed precision is not always feasible on major problems. The first issue is to determine when the assembly did come into existence, and this must carry one back into the seventh and sixth centuries.

I

Appearance of the Assembly

When Athens begins to be dimly visible in the mists of early Greece, its central government was very limited. Executive functions lay in the hands of nine archons, elected annually from the aristocrats by the council of the Areopagus. At least in later days the king archon (*basileus*) supervised the religious rites which bound together the community; the military commander (*polemarch*) led out his peers in time of border skirmishes; the archon proper, *archon eponymos,* who gave his name to the year, conducted the civil administration insofar as it existed; six "law-writers" (*thesmothetai*) were also added at an uncertain point. Their functions could not have been arduous in a period when few matters required decisions set down in the newly developed Greek alphabet.

To advise and check the archons the council of the Areopagus consisted of leading citizens, reenforced each year by the incorporation of the outgoing archons. As a standing body its powers were probably preeminent; generally in Greek states before 600 ancestral monarchy had been replaced, save at Sparta and Argos, by aristocratic councils drawn from the wealthier landlords.

There is no evidence at Athens in such early times for an assembly of citizens, but that is not to say that it could not have existed. The heroes at Troy as well as the inhabitants of Ithaca and the fairy-tale land of the Phaeacians in the *Odyssey* met when sum-moned by heralds to hearken to problems, but, as the story of the forward commoner Thersites in the *Iliad* suggests, they were not expected to vent their own views; in the Homeric epics assemblies were not structured and existed very largely to convey information. Auditors might shout or otherwise express their opinions, yet only when the Achaeans all began to rush toward their ships to aban-don the Trojan War can they be said to have made a decision, which was soon checked. A remarkably close parallel has been described in anthropological literature among the Tswana of Bechuanaland.[1]

At Athens citizens certainly came together for religious cere-monies; in other Greek states by the seventh century legislation of formal character began to appear. It is unclear who gave approval, for example, to the law at Dreros on Crete limiting the chief official to serving only once in 10 years, but we may postulate some form of public assent.[2] The one political event in early Athenian history on which there is any information was the attempt by Cylon and his fellow conspirators probably in 632 to seize control and estab-lish a tyranny, as had happened often elsewhere. The effort failed, and Cylon with his friends had to retreat to the Acropolis; to check them citizens were summoned "from the fields" in toto (*pan-demei*). When most of them went home they left the siege in the hands of the archons according to Thucydides; in discussing the Cylonian conspiracy Herodotus mentions instead the leaders of the naval districts called *naucraries*. This was more a levée en masse than a proper assembly.[3]

In the end Cylon escaped, but the others were tricked into leav-ing religious sanctuary and were arbitrarily executed by the archon Megacles of the famous Alcmeonid family, though in a few years all his relatives were temporarily expelled for the taint of sacri-lege, a charge which was to surface again in the days of his de-scendants Cleisthenes and Pericles. Oddly enough, the council of the Areopagus is not reported to have had any role at this juncture. Late in the seventh century a rule was passed forbidding any citi-zen from urging renewal of war against Megara over the possession of the offshore island of Salamis, a ban violated by the fiery young patriot and poet Solon, but the ordinance may have been enacted by the council of the Areopagus.[4]

In an excursus on early Athenian history Thucydides praises the legendary hero Theseus for uniting Attica (a process called *synoikismos*). Down to this time each local community had had its own officials, but from Theseus onward the plentitude of councils was supplanted by one council for all Athens.[5] It is indeed true that the Athenians very early showed remarkable political wisdom in fashioning the promontory of Attica into one state, the largest in mainland Greece, and in historical time celebrated a festival named the *Synoikia,* but that the institutions mentioned by Thucydides existed in early Attica may be doubted; whatever his merits in describing the vicissitudes of the Peloponnesian War he is an unsafe guide when he looked back into the past.[6]

The consolidation in historical reality was not as simple and conscious a step as Thucydides pictures it. On the east coast Marathon and adjacent communities, called the Tetrapolis, always remained distinct enough to send their own representatives to Delphic festivals; according to the Homeric Hymn to Demeter, Eleusis still had its own king when Demeter came there; frontier posts such as Eleutherae were annexed later as was Oropos on the north coast. Nor was Athens unique in this slow accretion. Argos destroyed Asine and expelled the inhabitants late in the eighth century and incorporated Mycenae much later; Corinth seized part of its neighbor, Megara, to improve control over the important isthmus of Corinth. The classical map of Greek states with immutable boundaries was achieved only slowly across the seventh and sixth centuries, a process as long protracted as the consolidation of the nation states of France, England, or Spain in the early modern period.

During the closing decades of the seventh century, troubles erupted in Athens, the consequence of political factionalism and also the exploitation of poor peasants by rich landlords who could even sell them into slavery for debt. So dangerous to the survival of the state were these issues that the well-to-do consented to the election of Solon as "archon and reconciler" in 594, the first reasonably firm date in Greek history. Solon, thanks to the preservation of much of his poetry, is the earliest figure in Athenian history whose reforms and motives can be explored with some confidence.

His far-reaching and astute economic measures, which were in the main successful, need notice here only inasmuch as the cancellation of rural debts and the ban on corporeal surety in the future gave the ordinary residents of the countryside an opportunity to assume an independent position which they might be able to exercise politically if given the opportunity. Solon's political reforms, on the other hand, while less effective in the short run, had a result which no one at the time could have foreseen, that of leading Athens on the path through the later reformers Cleisthenes and Ephialtes to the full democracy of the fifth century. Solon was more revered than any other person in Athenian history, a place which the first "leader of the people" (*prostates tou demou*) fully deserved in the breadth of his vision and skillful efforts to implement his revision of society and political structure.

Solon divided the citizens into four classes: *pentakosiomedimnoi, hippeis, zeugitai,* and *thetes,* on the basis of their agricultural revenues, 500 measures of grain or the equivalent, 300, 200, and less than 200 (when money came into use, the *thetes* were assessed as having less than 200 drachmae in property). Only the *pentakosiomedimnoi* could serve as archons; the three upper classes as a whole, who could afford the expense of armor, constituted the Athenian army in time of war. The important step was the admission of the *thetes* to the assembly, which now had certainly gained a role in public life. It was directed firmly, however, by the archons, who were appointed by lot out of candidates previously selected by each of the four tribes into which the Athenians were divided.[7] The assembly was thus more fully structured, though the only rule which Solon is said to have issued to govern proceedings regulated the conduct of speakers in its meetings.[8]

Deliberations were made more efficient by Solon's creation of a council of 400, consisting apparently of 100 members from each of the four tribes.[9] The only function of this council was to serve as a steering committee to screen business coming before the assembly (*probouleusis*). "Guardianship of the laws," that is, the protection of public order, continued to be in the hands of the council of the Areopagus. Among its powers were scrutiny of officials before they entered office (*dokimasia*) and holding them accountable (*euthyna*) for their actions; offenders could be punished and fined by it. These were fundamental levers in the control of Athenian government though they were not always to remain in the custody of the council of the Areopagus.[10]

From a sixth-century inscription it appears that Chios also had a "democratic" council of 50 members from each of its tribes alongside an older aristocratic council, and it was once thought that Solon drew his idea from Ionia;[11] redating of the Chiote inscription to 575–50 must make the influence flow in the other direction, if indeed there was any borrowing.[12] The view that Solon's model was the council of elders at Sparta seems equally unlikely.[13] All that one can be certain of is that new winds were blowing at several places in the Greek world toward the close of the seventh century and were weakening the grip of the local aristocracies.

Solon's other political reform was the creation of a law court; here too the *thetes* could serve, in the role of jurors. Anyone might sue on behalf of the wronged and appeal to the *heliaea* against improper actions of the archons. Both in Aristotle's *Politics* and in a late fourth-century monograph called *Constitution of the Athenians* (*Athenaion Politeia*), at times ascribed to Aristotle (though improbably), the *heliaea* is singled out as the cornerstone of the Solonian constitution and a mark of its democratic tendency.[14] The *heliaea* is often described as a judicial committee of the assembly, but even if its jurors were the citizens who also voted in the assembly, the two institutions are usually distinguished in the sources as separate entities. In the present work the judicial system of Athens need appear only when it functions in hearing appeals from the assembly or the council.[15]

Solon himself was no democrat, not that the concept had yet appeared. He mistrusted his fellow citizens as being too easily seduced by ambitious leaders and intended in all his reforms, both economic and political, to take a middle road of compromise. As he observed, "I gave the common folk [*demos*] such privilege as is sufficient for them, neither adding nor taking away . . . I stood with a strong shield thrown before the both sorts and would have neither to prevail unrighteously over the other."[16] Essentially his changes replaced a pure aristocracy by a timocracy in which the degree of public participation depended not on birth but on wealth; Aristotle analyzed the Solonian reforms as tripartite, "the council of the Areopagus standing for oligarchy, the method of electing the executive magistrates for aristocracy, and the system of popular law-courts for democracy."[17]

Political contentions especially for control of the post of *archon eponymos* continued to plague Athens after Solon had completed his mission. One man managed to hold the office for two years and two months; twice in the years after Solon no one was successfully elected to direct the state. The way was open for an ambitious aristocrat to gain power as tyrant, and Pisistratus proved to be the man who succeeded in subverting the established order, though only after three attempts.

His initial seizure of control in 561 lasted four years but was then halted by the opposition of his peers. During this first attempt, incidentally, the only known example of activity by the assembly in the early years of the sixth century took place when it approved a motion by Aristion that Pisistratus be allowed a bodyguard of club bearers; whether the council had first discussed the matter is not recorded, but evidently the later requirement that a decree be proposed by a specific person was already in effect.[18]

The second return by Pisistratus to power was very brief, and he had hastily to withdraw to Thrace where he owned gold mines. In 546 his third effort was much better based. This time he used force as embodied in mercenary warriors; he was also joined by "partisans from the capital and by numbers from the country districts" and won a decisive battle at Pallenis in east Attica.[19]

Once he was in firm control Pistratus sought deftly to gain wide support. He improved the water supply of the city by constructing an aqueduct, set off the marketplace (*agora*) by border markers, and built temples and other public structures as well as enhancing the majesty of the religious festivals of the state. For the rural population he provided loans and created a system of travelling judges so that farmers did not have to trudge to Athens for minor suits; on the other hand he instituted a levy of 5% or 10% on agricultural revenues, a very unusual step in ancient Greece where rural production was almost never taxed. He is reputed to have kept the constitution which he inherited formally in effect, so presumably the assembly and council of 400 continued to meet as usual though necessarily subservient to the wishes of the tyrant.[20] Archons were elected each year even if Pisistratus saw to it that his supporters gained the dignity of *archon eponymos*. As will appear in the next chapter, the council of 400 was certainly in existence immediately after the end of the tyranny, and the *heliaea* would have remained active, though it was before the council of the Areopagus, as was proper, that Pisistratus once appeared to defend himself against a charge of homicide—his accuser did not venture to press the issue.

Pisistratus died in 527; he had gained such popular favor that his reign was fondly remembered, according to Aristotle, as a golden age of Cronus. Two of his sons, Hippias and Hipparchus, succeeded peacefully and ruled jointly until Hipparchus was assassinated at the Panathenaic festival in 514 as the result of a personal feud. Harmodius and Aristogeiton were celebrated in song and later in statues as the "tyrant-slayers," but in reality Hippias continued to be master and exercised his control more despotically. A number of leading Athenians were driven into exile, including the Alcmeonid family, which fled to Delphi and there gained the contract to rebuild the burned temple of Apollo. They did so with great lavishness and suborned the famous oracle to support their efforts to oust Hippias by urging the Spartans to take action. At this time Sparta was the leading Greek state and had a reputation for disliking tyranny; in 510 it sent an expedition by sea under King Cleomenes to land at the Attic roadstead of Phalerum. Hippias was besieged on the Acropolis; after his children had been caught in trying to escape he agreed to evacuate Attica and sought Persian protection.

So once again Athenian aristocrats could freely resume their rivalries for public honors and prestige, as in those earlier years which had brought the election of Solon as archon. The rule of

Pisistratus and his sons, however, had seriously weakened their control of the countryside and had enhanced popular awareness of the unity of Athens; middling urban elements such as potters and tradesmen had also become more prosperous and independent-minded. This time the reaction to upper-class contentions was to be even more revolutionary.

II

Consolidation of the Assembly

After Athens had been liberated by the Spartans, the dominant politician initially was the aristocrat Isagoras, who wished to put back the clock by undoing the changes in the tyrannical period; one step toward this end would be the purging of aliens who had crept onto the citizen rolls. His main opponent was the Alcmeonid Cleisthenes, but in 508 Isagoras was elected *archon eponymos* to carry out his intentions. The ensuing tangled web of events is fairly well illuminated by Herodotus and the *Constitution of the Athenians,* though they do not agree on all points. It appears, however, that Isagoras may have been removed from office by Cleisthenes, who had "called to his aid the common people" with the battle cry of *isonomia* or equal rights.

As a counter Isagoras played the Spartan trump by inviting Cleomenes to return and restore the situation. The Spartan king came with 700 men and invoked the old "Alcmeonid curse" to force Cleisthenes to leave Attica. Isagoras and Cleomenes then proceeded to try to abolish the Solonian council of 400, probably to neutralize the assembly, but the Athenians rose in rage and besieged the Spartans and Isagoras on the Acropolis. After two days the Spartans were allowed to depart from Attic soil; Isagoras and his supporters were executed. Cleisthenes could now return and begin a remarkable set of reforms which may have required several years to draft and implement.

The Reforms of Cleisthenes[1]

His reorganization of the Athenian constitution required a very precise geographical knowledge of Attica and its social structure so that he could recast the voting districts to carry out his fundamental objectives.[2] Thenceforth Attica was divided into 10 tribes; each tribe in turn consisted of three districts (*trittyes*): one in and about the city, the other two in rural areas which might be contiguous but usually were separated by some distance. Each *tritty* in turn was composed of one or more demes, normally the preexisting villages of the Attic countryside; the demes in each *tritty* might be adjacent but this was not often the case. At least by the fourth century there were 140 demes.[3]

The purpose of this elaborate arrangement was to control the manner in which a new council of 500 was formed. Each tribe provided 50 members, a group eventually called a *prytany,* who were drawn by lot from candidates nominated by the demes. There is no information as to how these persons were chosen locally,[4] but in the fourth century it was possible for a man who wished to be a councillor in a specific year to engineer his appointment. At an unknown point the political year (but not the religious and civil year) was divided into 10 parts named after the tribe which provided the councillors, by lot, for each *prytany.* Decrees from the 460s on name the tribe in charge first, and its members had to be ready for immediate action in the council chamber if an emergency arose—indeed, one-third of the *prytany* ate and slept night and day in rotation at the *tholos.* To safeguard against the danger that the council could become a hidden master, like the Signoria in medieval Venice, councillors could not serve two years in a row or more than twice in their lifetime; as a result a very large proportion of Athenian voters, about one-third, would have taken a turn on this steering committee for the assembly.

By Cleisthenes' scheme, moreover, the council represented all areas of Attica; old geographical blocks, cult groups, and local control by the well-to-do were much weakened.[5] His boldness, to be sure, was not unprecedented. Reorganization of voting districts had earlier been carried through at Sicyon, Corinth, Miletus, and elsewhere; at Corinth the eight tribes were even divided into "thirds" of non-contiguous territory.[6] The Greeks were generally casting off inherited shackles of tradition.

The true aims of Cleisthenes remain the subject of debate, for there is no evidence to illuminate his basic attitudes, such as is provided in the poetry of Solon. Sceptical historians are inclined to suspect that Cleisthenes, in his grouping of demes into *trittys,* was seeking personal advantage, though it is difficult in looking at the new map of Attica to detect deliberate gerrymandering on a wide scale.[7] Clearly he did put the ultimate power of political decision in the hands of the assembly, though with at least one safeguard. The council of 500 thenceforth had to scrutinize every measure brought before the people and voted a preliminary proposal, *probouleuma,* which could contain either specific recommendations or present a more general statement so as to place an issue on the agenda of the assembly; without a *probouleuma* the people never acted on any matter. Another measure of security against rash decisions was to be introduced later, probably in the reforms of Ephialtes.

The social institutions of Athens were left untouched except in one respect. Earlier the basis for claiming citizenship had been acceptance of a son by his father's brotherhood (*phratry*), and a man was officially known as the son of a certain father. After Cleisthenes' reforms registration on the roll of a specific deme was the necessary step, and a citizen was formally called, for example, Themistocles of Phrearrhi, though patronyms actually continued to be widely used.

The safest conclusion is that Cleisthenes sought to gain popular support through the assembly, and equally important the mass of Athenian citizens had developed sufficient political consciousness that it was now able to exercise its new responsibilities;[8] immediately after the reforms the Athenians met and repelled an invasion by Chalcis and the Boeotians. As Herodotus comments on this victory, it proved "how noble a thing freedom is, not in one respect only, but in all; for while they were oppressed under a despotic government, they had no better success than any of their neighbors, yet, once the yoke was flung off, they proved the finest fighters in the world."[9] The Greek word here translated as "freedom" is actually *isegoria,* the right of equal speech, which appears to be a prerequisite for participation in the assembly, even if efforts have been made to show that this privilege was attained only in the middle of the fifth century.[10]

Apparently the reforms were all in place by 501/0 when the councillors swore an oath of office which continued to be used for the next two centuries, though with some changes; one significant clause directed them to prosecute anyone who subverted the democracy or aided an attempt at tyranny.[11] Athenians in the age of Cleisthenes were worried about the possible revival of tyranny, and with good reason. Hippias had been exiled, but he had the ear of the Persians and apparently lingering support in some Athenian circles; "the Athenians, with the customary leniency of the democracy, had permitted the supporters of the tyrants to remain in the city, as far as they had not personally taken part in their wrongs at the time of the civil disorder."[12] In 490 Hippias did return with the Persian expedition sent by Darius to avenge Athenian support to the Ionian rebels; after the amazing victory of the Athenian army at Marathon it marched back as fast as possible to the city, partly to counter any effort at a plot by a Pisistratid faction.

As a permanent measure against would-be tyrants Cleisthenes instituted the system of ostracism. On the agenda for the sixth prytany each year was a vote by the assembly whether to hold an ostracism; if it so decided the citizens lined up on a set day in the Agora and secured potsherds (*ostraka*) inscribed with the name of a politician whom they disliked or feared. For a verdict there had to be a minimum of 6000 sherds in the urns; this was the only occasion on which citizen votes were formally counted. The man who received the most votes was duly punished by exile for 10 years though he did not lose his citizen rights.

Some scholars point out that ostracism did not come into play until 487, and doubt that Cleisthenes was the author of this safety-valve, but it bears the stamp of his shrewd ingenuity. Initially, as just suggested, it may have been intended as another check against tyrannical elements; some surviving *ostraka* thus bear curses against a possible victim as Medizing. Ostracism, however, could have a wider purpose in Athenian public life. Many years earlier Periander, tyrant of Corinth, had asked the advice of his fellow tyrant Thrasybulus of Miletus as to how to deal with malcontents; Thrasybulus mystified the messenger by simply walking through a wheatfield and snapping off with a whip stalks which stood too high. Periander, however, understood the message, and Athenians of noble birth who became too ambitious might properly be wary of the danger of being ostracized as threatening "the national consensus, especially by publicly advocating ideas or acting in ways that threatened the values of political society."[13]

Solon's name was often invoked by fourth-century Athenian orators, and at least from the time of Plato he was considered one of the Seven Wise Men, ethical and philosophical guides of early Greece given to aphorisms; the terms "solon, solonisch" appear indeed in modern political vocabularies. Cleisthenes enjoyed no such fame. After his reforms he completely dropped out of sight, but since Cleisthenes was probably born several years before 560 (he was archon in 525/4) he may simply have died around 500.[14] In later years he was almost forgotten and never became a figure of reverence even though he, more than any other man, codified the Athenian processes of reaching political decisions which remained in force for the next two centuries, apart from the brief upheavals of 411 and 404. His reforms perhaps smacked too much of opportunism, as Herodotus tacitly suggests in describing his maneuvers against Isagoras. "He did not create the democracy but made it possible," and further steps were to be necessary before Athens can be said to have advanced into full democracy.[15] Yet more than once in history opportunists have had greater effects than they planned in seeking their own advantage; now that the rule of the assembly was consolidated it asserted itself ever more.

Developments of the Early Fifth Century

New politicians took the place of Cleisthenes during the first decades of the fifth century; in Plutarch's lives they are pictured as Aristides, a conservative leader, and the more radical Themistocles, though in the *Constitution of the Athenians* Aristides appears to have collaborated on occasion with his purported opponent.[16] There is, however, no indication that any significant figure of the era wished to subvert the new democracy,[17] and the assembly to which they had to appeal for support begins to be slightly more visible in literary and epigraphic sources.

When the Ionians were meditating rebellion against the Persian overlords in 499 they sent the Milesian leader Aristagoras to mainland Greece to seek aid. Naturally he went first to Sparta, the balance wheel of the Greek state system, but there he was unable to win over King Cleomenes. At Athens he gained a far more favorable reception from the council, which first heard ambassadors, and easily persuaded the assembly itself to send temporary aid in the form of 20 ships. As Herodotus dryly observed, "It seems indeed easier to deceive a multitude than one man."[18]

Already in the first year after the end of the tyranny the assembly had passed a decree banning the use of torture, but this is known only from a literary reference.[19] Now decrees inscribed on stone began to appear. The first, which seems to be from the last years of the sixth century, regulates the duties of settlers on the island of Salamis, finally wrested from Megara, and begins simply, "the people decreed" (*edoxen toi demoi*); the council does seem to be referred to in the last line as exercising its probouleutic function, but "there is no certainty of this."[20] The second surviving example, dated to 484, consists of two inscriptions concerning the temple of the Hecatompedon; once again the *demos* is twice noted as passing the decrees.[21] From 480 there purports to be a decree moved by Themistocles ordering the evacuation of Attica during the Persian invasion, but the authenticity of the "Themistocles Decree," which survives in a third-century inscription from Troezen, is more than debatable.[22] Beyond question, however, the assembly now considered its actions worthy of being published in permanent form on steles.

The assembly also took to itself other powers, first of electing the 10 generals by open vote, but on the other hand shifting the manner of choosing archons from vote to choice by lot from 100 or 500 candidates nominated by the demes.[23] It has been argued that this latter step weakened the prestige of the council of the Areopagus, into which the archons passed on completing their year of service, but we have no evidence as to the character of men proposed by the demes; after all they were still drawn from the upper three Solonian classes.[24]

By this time the machine of ostracism had been put into action, first in 487 against the aristocrat Megacles and other erstwhile supporters of tyranny; very soon Themistocles directed the attacks against his own opponents, ending in 482 with the ostracism of Aristides. As a result he was to be the unquestioned leader of Athens when it came to face the onslaught of Persian forces by land and sea under King Xerxes.

The Persian Invasion

Here we must turn aside briefly to consider in a wider light the course of the invasion and its almost unbelievable repulse; to repeat a comment in my Preface, Athenian political development took place within the context of Greek history as a whole.

In 490 the expedition which the Persians launched by sea against Athens by sea had been defeated on the plain of Marathon, but King Darius died too soon thereafter to redress the situation; his son and heir Xerxes had to consolidate his position for several years both internally and externally. Eventually he decided to prepare a full-scale invasion under his own command to conquer all the Greek mainland, and duly marshalled overwhelming forces by land and sea in the fall and winter of 481/0 in Asia Minor.

Most Greek states and their leaders had been blind to this looming danger, but there was one man, Themistocles, who saw the threat and also was able to fashion a weapon which was to be decisive in giving the Greeks a chance to meet the Persian attack. In 483/2 an unexpected and rich strike was made in the Athenian state silver mines of Laurium; normally such revenues were divided among the citizens, but Themistocles persuaded the assembly to divert the funds into building warships and begin the fortification of the promontory of Piraeus to provide a better port than the open roadstead of Phalerum. To his fellow citizens he advanced a justification in terms of Athens' inveterate enmity to the neighboring island of Aegina, but his vision was far wider.

Almost all of the Greek states shuddered in dread as the threat became more obvious and either submitted to Persian envoys or remained neutral; on the tripod which was erected at Delphi after the final victory only 31 *poleis* were named out of the hundreds in the Aegean and overseas.[25] Those who decided to resist met at the isthmus of Corinth in the fall of 481 and made a vital decision; taking note of the fact that the Ionian rebels had failed to establish a single command, they agreed that the Spartans should provide leadership by land and also by sea.

There was no question that the Spartans, who had the best trained and largest band of hoplites, should have preeminence on land. In regard to naval command the Athenians, who with their newly built ships provided two-thirds of the Greek navy, could have been stubborn, as an Athenian ambassador baldly told Gelon, tyrant of Syracuse, in an unsuccessful effort to gain his aid; but to secure unity Themistocles surrendered any claim to furnish admirals.

Whether the Athenian assembly raised any objections when he reported this wise step is unknown, but it did accept his judgment on a related matter which is briefly reported in Herodotus, the earliest known formal debate of the citizens.[26] Earlier the Athenians had sent envoys to determine the counsel of Apollo at Delphi. The god's first prophecy was one of total disaster; on further appeal to the oracle a softer answer promised that a "wooden wall" would protect the Athenians and their children and spoke in riddling language of a battle at Holy Salamis. In the meeting of the assembly convened to meditate on Apollo's advice opinion was sharply divided between elder citizens who wished to rely on the Acropolis, once protected by a palisade, and others who argued that the god meant the ships which "had best be got ready." Doubts as to the meaning of the reference to Salamis persisted but were set aside by Themistocles, who pointed out that Apollo had described Salamis as "holy" rather than "unlucky" so that if they chanced to fight on the sea at Salamis they would win. Since the Athenians stood behind Themistocles throughout the dire events of 480–479, they accepted his deft interpretation and prepared to do battle if necessary by sea.

On the Greek side the basic strategic decisions in this and other problems were the fruit of Themistocles' incisive, keen judgment and his persuasive tongue, backed by the knowledge of the other leaders that he had the support of his fellow Athenians. He had put his finger on the critical flaw in the Persian attack, the fact that the Persian fleet must remain close to the coast both to protect the seaborne supplies of grain for the vast army and also to be safeguarded itself when it landed its weary rowers on shore at night. Marathon had shown that Greek hoplites had a chance against the more lightly armored Persian infantry, but how could the allies meet the Persian superiority at sea in numbers and in skill, thanks to the Phoenician contingent?

The solution lay in enticing the Persian navy into a narrow body of water where its strength could not be fully deployed. The first site the Greeks chose was the strait of Artemisium between the island of Euboea and the Boeotian coast, but this was outflanked when Xerxes' army was able to force the pass of Thermopylae. The Greek fleet then fell back to the island of Salamis off Attica.

The Athenians, men, women, and children, evacuated their homeland as the Persian army rolled down through central Greece. Even if one may doubt the validity of the "Themistocles Decree," there can be no question that this manifestation of desperate determination was set in motion by a decree of the assembly.[27] To secure unity the citizens also revoked all earlier decrees of ostracism, and the council of the Areopagus distributed the state treasury by giving each man eight drachmas (the standard coin was a four-drachma piece) for his support on Salamis or the Peloponnesus. Herodotus was right in his judgment that "it was the Athenians who—after God—drove back the Persian king."[28]

The council of Greek leaders at Salamis fell to wrangling, and some wished to retreat to the isthmus of Corinth, where the Peloponnesians had been busy across the summer building a defensive wall; but Themistocles strongly disagreed. In open waters the Greeks still could have little hope for victory, and Argos, which lay behind the wall, was neutral out of opposition to its old enemy, Sparta, and would certainly join the Persians if given a safe opportunity.

His colleagues finally agreed that if he could entice Xerxes into committing the Persian fleet in the narrow waters of the strait off Salamis they would stand and fight. By wily messages to the Persians, suggesting that the Athenians were ready to give up, Themistocles was successful, and in the ensuing battle some 200 Persian warships, mostly of the Phoenician cities, were lost as against only about 40 Greek ships. Xerxes left behind part of his army to try to win by land, but in 479 the hoplites of Sparta, Athens, and a few other states crushed the Persians at Plataea.[29] Themistocles had been feted at Sparta as the architect of victory, but already at Athens his leadership of the assembly was weakening even though the citizens did approve his proposal to build a wall hastily around the city. Athenians did not easily endure the overweening pride which he displayed.

Internal Developments after the Persian Defeat

In Thucydides' judgment it was Themistocles who first convinced the Athenians that their future lay on the sea. Across the sixth century Attic exports of vases had swept all Mediterranean markets, but as a state Athens played a smaller role on the sea than did Aegina, Corinth, or for that matter Sparta. Much later a Syracusan leader was to point out to his fellow ctiizens that "the Athenians were more landsmen than the Syracusans and had only taken to sea when forced to do so by the Persians."[30]

According to a generally accepted opinion, this change indirectly strengthened the political position of the *thetes,* who manned the navy and also benefited from the growing commercial activity of the Piraeus. Any such interpretation must be viewed with caution for two reasons. First, much of the trade and industry of Athens was in the hands of resident aliens (*metics*) such as Cephalus, in whose garden Plato laid his *Republic,* who did not necessarily wish to share the heavy burdens of citizens,[31] and second the *thetes* themselves, as we shall see more fully in the next chapter, did not often carry much weight in the assembly.

The main consequence of the Persian Wars was a major reenforcement of the self-assurance of the Athenian citizens. It had been a vote of the assembly which led to Athenian aid to the Ionian rebels in 499; thereafter the creation of a large navy and the decisions to stand firmly behind the Greek resistance to the Persians were taken, as far as we know, with no major opposition.

In 479, indeed, the Persians sent two envoys to Athens urging it to surrender; the second, Murychides, came first before the council of 500, one member of which suggested that he be given a hearing in the assembly. His fellow councillors and the Athenians cut short the proposal by stoning him and his family to death, and the Athenians remained loyal to the Greek alliance.[32]

Immediately after the Persian retreat there were two major shifts in Athenian history. The first was the unconscious beginnings of Athenian empire; the second, a conservative revival under the leadership of Aristides and Cimon.[33] Aristides, famed for his probity, was the Athenian representative in the creation of the Delian league in 478/7, which joined the strengths of islanders and states on the coast of Asia Minor together with Athens to prevent a return of the Persians; he was also given the delicate job of assessing how much each state should contribute in ships or, for smaller communities, cash to support naval operations. By 471 the opponents of Themistocles, who had on the evidence of surviving *ostraka* often voted against him in the 480s, were strong enough to engineer his own ostracism; he was never able to return to Athens and died a pensioner of the Persian king.

An important factor in the conservative dominance in the 470s and 460s was the trust of the Athenian citizens in Cimon, son of that Miltiades who had led the Athenians at Marathon; for more than a decade he was the major leader of the community. Year after year he was elected general, perhaps the greatest field commander Athens ever produced, and swept the Persians out of the Aegean; his uninterrupted successes culminated in a smashing victory over the renascent Persian fleet at the Eurymedon river on the south coast of Asia Minor (probably about 465). The booty provided the funds to build the great south wall of the Acropolis, which still stands today; Cimon himself was renowned for his liberality from his estates to his fellow demesmen. Cimon also favored friendship with Sparta; when its helots revolted and ambassadors begged Athenian aid, he secured approval from the assembly for a force to give support to its "yoke fellows"—but unfortunately the Spartans changed their mind and rejected the assistance.[34]

The Beginnings of Athenian Empire

Aristides had had a major role in drawing up the charter of the Delian league, by which Athens, as contributing the greater part of the league navy, was allowed to appoint its generals and the treasurers of the funds stored at the temple of Apollo on Delos. Initially the other members of the league probably gave little thought to these provisions, but eventually they were to regret them deeply.

During the years of Cimon's leadership the nature of the bond subtly altered. When the small state of Carystus on the island of Euboea was liberated, it was forced against its will to join the league; the crusade must not be weakened by local unwillingness to participate. Then the island of Naxos grew weary of the annual burden of providing ships; Athens could not tolerate any effort to avoid the common duty. Worst of all was the "revolt" of Thasos, a large state, which had to be recalled to its fealty by a siege. Since public documents before the middle of the fifth century have rarely survived, we cannot be sure how far Cimon's activities called for formal endorsement. In later decades the settlement of local dissidence seems usually to have given rise to an Athenian decree, examples of which will be noted when we come to Pericles, but in earlier years the assembly of the League may still have been active enough to give its approval for the expenses connected with the Thasian siege.

Even so, voluntary league was slowly, almost unconsciously, converted into empire under the sole direction of Athens. Modern students date the point at which the process was complete to 454, when the treasury of the league was moved from Delos to Athens, where Athena and her priests could better protect it on the Acropolis.[35]

The Reforms of Ephialtes

By the late 460s the pendulum of Athenian leadership swung back from conservatives to far more radical figures, and remained in their hands down well into the Peloponnesian War. The decisive change of the Athenian voters in this shift was the ostracism of Cimon in 461.

The ultimate causes for the reversal are not even dimly illuminated in the scanty sources; as is often the case in ancient history one must look at subsequent events to gain guidance on the forces which stamped them. In the present case it can safely be said that Ephialtes and then Pericles advanced policies which suited the changed temper of the era: hostility or at least suspicion of Sparta; open imperialism; and the removal of conservative trammels in the structure of Athenian government. These were to be the principles to which the assembly adhered, rejecting all conservative efforts to check their implementation.

Ephialtes was at once one of the most significant reformers of the Athenian constitution and also its most shadowy figure.[36] The limited sources agree that during the Persian war the council of the Areopagus had secured wider prestige in its ancestral role as "guardian of the laws"; in discussing changes in constitutions Aristotle noted that "the council of the Areopagus, for example, gained in reputation during the Persian war, and the result appeared for a time to be a tightening of the constitution [i.e., a movement in the direction of oligarchy]." Then the tide turned, and in 462/1 Ephialtes, after attacking several Areopagites on charges of administrative misconduct, obtained legislation which severely clipped the powers of the Areopagus and "plunged the city into unmitigated democracy."[37] Thereafter the Areopagus only conducted trials for homicide, poison, arson, and minor matters.[38]

What jurisdiction had it lost? The answer depends on the functions which the Areopagus still exercised in 462, but this is a problem which cannot easily be solved in view of the uncertainty which attends Cleisthenes' reforms, possible changes in the early fifth century, and the lack of clear, relevant evidence. Recently it has been argued by several scholars that the Areopagus was deprived of three decisive controls: the right to institute charges of treason or malfeasance (*eisangelia*), scrutiny of incoming magistrates (*dokimasia*), and auditing of their performance (*euthyna*).[39] Definitely the Areopagus had had these functions in the sixth century, but it did not in the later fifth century.

The interesting point is where these powers were then lodged—not in the assembly, but in the first instance in the hands of the council of 500. Its scrutiny of appointments by lot as well as the councillors for the next year was largely a formal determination of ownership of land in Attica and other qualifications, but presumably it could reject a man physically or intellectually incapable.[40] For financial checks the council relied on boards of experts who referred offenders to the jury courts. Its right to hear accusations of *eisangelia,* directed mainly at generals and politicians, "put the disposition of crimes against the state entirely in the hands of popular organs," though the council could fine only up to 500 drachmas; other cases had to be laid before the juries or less often the assembly.[41] Another change may have had a significance which has never been heavily stressed. In early decrees, as already noted, "the people decided"; in decrees passed after the time of Ephialtes the prescript usually read, "the council and the people decided" (*edoxen tei boulei kai toi demoi*). The role of the council as a regulator of public debate may thus have been formally enlarged and the efficiency with which state affairs were conducted much improved.

As far as the assembly itself was concerned, there is no real evidence of transfer of powers beyond a bland statement in our sources. Indeed, since the days of Grote there has been a rather general consensus that Ephialtes may have been responsible for a further check on rash action in the assembly by the introduction of the *graphe paranomon,* "writ of unconstitutionality." This view has sometimes been questioned inasmuch as the first recorded use of the *graphe* comes only in 415. Our detailed information for events in the fifth century, however, is so spotty that this can hardly be taken as decisive, and there seems offhand no reason why the writ should have been introduced in 415. Rather it can be pointed out that Athenian reformers from Solon and Cleisthenes onward often coupled an increase in popular jurisdiction in the assembly with safeguards against its misuse, and Ephialtes might well have followed this principle. In sum, one may follow Grote and other historians only with caution in assigning the *graphe paranomon* to the period of Ephialtes, but certainly this safeguard against faulty decisions was well in place before the close of the fifth century.[42]

Every decree passed by the assembly had to be proposed by a specific, named individual, who was, however, almost never a political figure of importance (save in the case of the Themistocles

Decree)—and with good reason. If later meditation decided that a decree had been improper after the introduction of the *graphe paranomon,* the formal initiator could be sued in the law courts. Upon conviction within one year the decree was invalidated, and the proposer was subject to a very heavy fine; after a year only the decree could be revoked. Between the requirement that the council provide a *probouleuma* on every issue laid before the assembly and the possibility that a charge of *graphe paranomon* might ensue later the assembly's deliberations were channeled with care.

Ephialtes has also been considered, though again without real attestation, to have divided the *heliaea* into jury panels (*dikasteria*) to accommodate that Athenian enthusiasm for lawsuits which is satirized at the beginning of Aristophanes' *Birds;* in Plato's *Apology* Socrates points out that he has never been in a law court, though this may be no more than a rhetorical ploy. Eventually the rules by which jurors were assigned to their panels were ingeniously codified so that no one could know before the day of a trial who would hear a case; the Athenians believed wholeheartedly in democracy, but neither in the legislative nor in the judicial sphere did they trust individuals.[43]

In the year in which Cimon was ostracized, 461, Ephialtes was stabbed in the night by an unknown assassin; murder of politicians was far less common in Athens than in Rome of the Late Republic, so evidently there were those who deeply resented his reforms, which had replaced "the paternalism of the aristocratic state" by "the open, permissive society of democratic Athens."[44] His assistant and then successor as leader of the radical wing in Athenian politics was Pericles.

The Age of Pericles

Pericles was of the bluest blood, his father the aristocrat Xanthippus, his mother Agariste, a niece of Cleisthenes, but he early showed popular tendencies by joining in a prosecution of Cimon in 463 and then in aiding Ephialtes.[45] After the death of the latter, he is repoted to have become *prostates tou demou,* though his role is not very clear until about 455, when he was a general. By 451 he was incontestably the leading figure in Athenian politics and in that year secured the decree that male citizens must be the off-spring of Athenian females as well as males.[46] Alongside this emphasis on the exclusive, majestic quality of citizenship, which was a *leitmotif* in Pericles' internal policies, went a far more open exploitation of Athenian power as master of a seabased empire. The assembly thus arrogated to itself the authority to pass a decree to tap the surplus funds of the Delian league in order to begin the embellishment of the Acropolis; this led at once to the creation of the Parthenon, the most expensive temple ever erected in the Greek world.

Not all citizens favored this and other measures. The conservative well-to-do, called *kaloi kagathoi* (handsome and good) had supported Cimon; now they turned to Thucydides, son of Melesias (not the famous historian).[47] In 445 or 443 Thucydides challenged Pericles to an ostracism contest on various charges, the only one of which we know is the assertion that Pericles had made the Athenians into harlots, adorning themselves with the revenues of empire.[48] These were coldly rational arguments not likely to appeal to the Athenian voter. The passion driving them, which Pericles skillfully exploited then and in later years, was in the words of Jacqueline de Romilly, "the desire which they have for fame, renown and honours. . . . In its highest form, their ambition aims at glory, in its lowest at the use of power."[49]

Thucydides lost the contest. Under Pericles' guidance the assemble legislated so as to interfere more and more openly in the autonomy of the erstwhile allies, now really subjects, by decrees ordering the use of Athenian weights, measures, and coinage or imposing democracy on states which dared to rebel.[50]

Probably late in the 450s, for example, there was dissension at Erythrae, a state on the coast of Asia Minor, perhaps caused by an element which sought to turn to the nearby Persians. The Athenian assembly ordered that it set up a council of 120 like the Athenian council, the members of which were to swear oaths of loyalty to Erythaean democracy and Athenian democracy; there were to be no more political expulsions or recalls of exiles without the authority of Athens.[51] Later Chalcis on the island of Euboea tried to throw off its yoke and was forcibly recalled to allegiance.

In a decree moved in the Athenian assembly a rider was added that while Chalcis might try its own citizens, it could not inflict "exile, death or loss of citizen rights. In regard to these appeal shall lie in Athens, in the law court of the Thesmothetae, in accordance with the decree of the people." As an anonymous critic of Athenian democracy in the 430s called Pseudo-Xenophon or the Old Oligarch dryly observed, this was good for the Athenian hotel business; it also safeguarded friends of Athens from local mistreatment.[52]

The power of Pericles, to be sure, continued to be contested but only indirectly. Phidias was charged in 438 and condemned for embezzlement of the funds for the great gold and ivory statue of Athena, and another friend of Pericles, the philosopher Anaxagoras, had to leave Athens hastily before being attacked on grounds of impiety for asserting that the sun was a molten rock as big as the Peloponnesus.[53]

From at least 445 Pericles was reelected general year after year until his death in 429. As his great admirer, the historian Thucydides, put it, "Athens, though still in name a democracy, was in fact ruled by her first citizen."[54] Yet in 429 the people could round on him, angered by the privations of the war against Sparta, and temporarily remove him from office as well as levy a fine. The assembly, one must never forget, was the source of ultimate decisions in government.

Modern students have almost always accepted Thucydides' judgment. "Pericles' long eminence was due to his incorruptible character, a consistently intelligent policy, and remarkable powers as an orator."[55] He not only led his fellow citizens into unquestioned acceptance of empire but also had a lofty view of the necessity of uplifting them culturally, a view well expressed in the Funeral Oration which Thucydides had him pronounce as a solemn commemoration of the men who died in the first year of the Peloponnesian War.[56]

Pericles was Thucydides' hero, and the historian sought to distinguish between his thoughtful political and military policies and the headlong abandonment of caution by his successors. Yet Plutarch, Thucydides, and the comic poets are agreed that Pericles' will was sovereign in Athenian political life after 450,[57] and it was in the succeeding decades that the decrees which consolidated Athenian mastery of its empire were enacted at the expense of violating the fundamental premise of Greek state policy, the right of each community to use its own laws (*autonomia*).[58] Pericles must also bear a heavy responsibility in the steps which led to the outbreak of the Peloponnesian War; he instigated the decree banning Megarians from the marketplaces and harbors of the empire and other affronts to the strength and pride of the Spartan allies, who were barely able to persuade Sparta itself to take up their cause. Of all ancient figures, in sum, Pericles was perhaps the most devastating influence on his state, first by helping to set off a war which Athens could never win but only hope to draw and then by already having led the Athenian assembly into that open imperialism which made inevitable the eventual collapse of its rule.

The Peloponnesian War

This grueling contest had shattering effects in the long run on the Athenian citizen body and on its decisions both in the assembly and in the law courts, but first it will be useful to sketch briefly the actual course of the war as a background for later discussion of some of the distortions of justice caused by its stresses. Conventionally the struggle with Sparta is dated 431–404, but it consisted of two very different duels between Athens and Sparta.[59]

In the first phase, 431–21, the Athenians generally followed the cautious policy which Pericles had laid down and wore down the Spartans and their allies by naval operations around the Peloponnesus. By 421 the Spartans were willing to make a peace, which was to last 50 years; in doing so they sacrificed the interest of Corinth and other states.

The Athenians, however, were so dissatisfied by paucity of results from their exertions and the loss of manpower in a great plague that they yielded eager ears to the blandishment of the popular leader Alcibiades and voted a great expedition to Sicily in 415; Nicias, a conservative, urged in vain the necessity of keeping their strength in the Aegean intact, and his assessment of the dangers of the Sicilian gamble only led the assembly to vote an increased commitment of effort. The amphibious operation, one of the largest in antiquity, had come to a disastrous end by 413, with the loss of all the ships and troops involved.

By that time Sparta had renewed warfare and could count on support from the disaffected Athenian subjects as well as financial aid from Persia. The Athenians fought on doggedly, but the assembly became unsure of itself and displayed too often an unwonted ruthlessness, examples of which will be treated in Chapter IV.

In 405 the able Spartan admiral Lysander swooped across the Hellespont and seized the Athenian ships beached at Aegospotami while their crews were on shore. He then drove Athenian settlers in the Aegean into Athens, which hopelessly endured a siege into 404, when it had to surrender unconditionally. Lysander allowed it to keep 12 ships but tore down the Long Walls which had safeguarded Athenian communications with the Piraeus; flute girls played to celebrate the joyous occasion. At Athens an oligarchic regime was given full authority; the erstwhile subjects quickly discovered that they had simply exchanged Athenian mastery for Spartan rule. That domination was to be far more erratic in its application and brief in duration both at Athens and in the Aegean generally.

III

Voters in the Assembly

If we are to understand how Athenian public life was conducted both during the peaceful years of the fifth century and also in reaction to the stresses of war we must look more closely at the composition of the assembly itself as fashioned by the reforms of Cleisthenes and Ephialtes into the rudder of the state and thereafter at the variety of its functions and its mode of operations. Even in the matter of attendance at the assembly there are two distinct, knotty problems which require thoughtful analysis.

Who Could Vote?

At first sight the answer to this query is simple and may be summed up in one phrase: all male citizens who were 18 years and over. In reality the issue is not so straightforward and leads directly into one of the most serious charges against Athenian democracy.

Sons were presented at the proper age by fathers to their demesmen: two oaths had to be sworn—that the son was 18 and that he was truly a citizen, that is, the product of a legal marriage or had been officially adopted to carry on the family name. Marriage was essentially a private matter, conducted according to ancestral customs;[1] but in 451/0 Pericles carried the rule that both parents had to be Athenian citizens. Earlier this limitation would have prevented Themistocles, Cimon, and others from any public role; and later it was to cause trouble to Pericles himself inasmuch as his son was born to his mistress Aspasia, a Milesian, and had to be given full citizenship by a special decree. Resident aliens could be enfranchised by the assembly, but this was more unusual in the fifth than in the fourth century; Themistocles could only reward his pedagogue Sicinnus, who played a vital role before the battle of Salamis by taking false messages to Xerxes, by buying him citizenship in Thespiae, not Athens. If there was any challenge to the right of the son to be registered on the deme rolls, demesmen were appointed to prosecute the issue before the jury courts.

At its height before the Peloponnesian War began to wreak havoc, the voting population has been estimated at 43,000. To this figure one must add an equal number of women, children under 18 (who would probably have been as numerous as adult men and women together in light of demographic principles applicable generally to preindustrial populations in the modern and ancient worlds alike), and perhaps 60,000 slaves, though reserves on this latter figure will be presented later. Of the population of Athens (172,000 citizens plus 60,000 slaves) about 18.5% could have been voters.[2]

This does not look like a very democratic structure. Athens has consequently often been criticized in recent years as being "a fraud, and the citizens were a leisured minority living on the labour of slaves"—a view which neatly ties together two current abominations, elites and slaves.[3]

Such a line of attack, however, will not do. Athenian slavery will be more fully assessed at the end of this chapter, but it may be observed that in all parts of the globe slave labor has been an endemic institution in societies which reached the level of civilization. Ancient slavery, moreover, has received far too much attention of late; most men clearly owned no slaves. The fact that women could not vote must not mislead us either. Only in the twentieth century have females been widely enfranchised, and certainly in ancient Greece, where they usually had no property rights, they were not deemed able to exercise independent judgment; at Athens they could enter a court of law only through a male spokesman.[4] Democracy, after all, is a system of government in which those who are generally considered capable of assessing political issues for themselves have the ultimate voice; the extent to which they choose to vote is incidental, not an essential criterion.

To support this view let us turn to a modern example, the United States presidential election of 1860. Years before Jacksonian democracy had removed restrictions based on ownership of property and other limitations on popular sovereignty; in the election itself a man who had been only a minor Illinois lawyer was chosen as president. Surely we can call the American political system democratic by that time.

The census returns for 1860 report 31,443,321 residents in the United States, of whom about 4,500,000 were identified as Negro, mostly but not entirely slaves.[5] The whites were almost evenly split between male and female, but since over half of the population, like ancient Athens, was still in childhood there were probably only about 6,300,000 adults who could have voted. From this total one should deduct the small number of inhabitants of the territories, who had no voice in the American electoral college, and a somewhat larger body of recent immigrants, though some states allowed those who were in process of naturalization to vote.[6] A limiting factor of greater weight was the rather strict requirement of length of residence before one could be registered to vote. Since this reduction cannot be quantified we may return with caution to the figure of 6,300,000; the conclusion must be that at the outside 20% of the population of the United States in 1860 could have voted. This is not far off from the estimate given above that 18.5% of the Athenians had the right to participate in the assembly. Actually in the election of 1860 there were 4,700,000 votes.[7]

In sum, eligible voters in ancient Athens were no more a leisured minority than those Americans who could exercise their franchise in 1860. The percentage who did cast ballots in presidential elections, indeed, was to sink in subsequent years; from 1952 to 1976 only 61.4% of possible electors voted, and the proportion has dropped even further more recently though American citizens would by and large indignantly deny the charge that their government was not a democracy.[8] It may also be noted in comparison that voting in modern America is normally a simple procedure which takes very little time; in Athens the exercise of political rights demanded far more attention.

Who Did Vote?

Those who could vote in fifth-century Athens had to be physically present in the assembly when it met in order to express their judgment—this was a direct, not a representative form of government. The rearrangement of the hillside of the Pnyx, where the assembly usually convened, in 460–400 would have given room for about 6000 citizens, the total required for an ostracism and some other votes.[9] In the early decades of the fifth century the number of citizens was much less than the 43,000 suggested for the high point before the Peloponnesian War; there may have been some 30,000 citizens of whom two-thirds were *thetes* dwelling mainly in the countryside. So the expectation that about 6000 voters were about all that could be assembled from the city and nearby rural districts may have been reasonable;[10] by 431 this figure represented a low minority of the citizen body. During the war itself, according to Thucydides, there were never even 5000 attendants,[11] though at some point in the years 418–15 there must have actually been 6000 to carry out the ostracism of the insignificant politician Hyperbolus, a farce produced when the rival leaders Nicias and Alcibiades joined forces; ostracism was never again used at Athens.

Why was attendance so limited, and who really did come to meetings of the assembly? Down to the Peloponnesian War, as Thucydides informs us, most citizens lived in the rural areas, and various factors reduced the likelihood of their participation in Athenian govenment. Geographically the distance to the farthest boundaries of Attica was about 25 miles; "the journey on foot or mounted on a donkey must have taken a full day."[12] This was not just a problem in ancient times; after the first election under the new American constitution in 1788 the Virginia legislature complained that many could not vote "by the necessity of traveling nearly fifty miles, passing over many mountains and crossing many watercourses (frequently so high at the time of holding elections as to make it dangerous to pass the same)."[13] Athenians did not have to worry about high water, but unlike Virginia electors they were unlikely to come by horse, a luxury reserved for aristocrats in ancient Greece. Even those who could afford a donkey were among the richer part of the rural population.

Another difficulty for Athenian farmers lay in the demands of the agricultural year, especially at the times of plowing and reaping harvests. Those who did the physical work on the land certainly could not attend 40 meetings of the assembly in a year or even a major proportion;[14] only a very important issue was likely to draw *thetes* as against the wealthier citizens who had hoplite status and did not necessarily have to labor themselves in the fields.[15] Yet it does remain true that references in Aristophanes and elsewhere attest that at least some farmers (*agroikoi*) did form a portion of the attendance at assemblies.

By 431 the inhabitation of Athens and the Piraeus had swelled greatly; Gomme estimates that about 15,000 potential voters then lived in town.[16] As he and others have noted, most of the political leaders of the fifth century already belonged to urban demes, but there is a strong tendency in modern scholarship to assume that the *demos* now consisted largely of *thetes* who dwelt within the city walls.[17] To support this assumption one might rely on the comment in Xenophon's *Memorabilia* that the voters in the assembly were fullers, shoemakers, carpenters, smiths, peasants, and retailers, comfortable to the description in Aristotle's *Politics* of the radical form of democracy in which the *thetes* had the main voice.[18]

Yet it should be remembered that Aristotle's survey reflects conditions in Athens in his own day, when attendants at the assembly were paid a modest sum; in the age of Pericles there was no such reward. Once again the view that "the citizens were a leisured minority" has distorted our picture of economic realities in fifth-century Athens; most citizens had to work themselves and often in grueling labor at the docks or in the city. Even Xenophon's description suggests that what may be called a middle class was most likely as a rule to attend the assembly; so too the citizens whose clothes were filched by Praxagora and her supporters (in the *Ecclesiazusae*) so they could participate in a meeting seem to have been solid folk.

The glowing picture in Pericles' Funeral Oration of the willingness of all citizens to bear the burdens of government accords ill with the simple fact that the Pnyx in his day was designed to accommodate at the most about one-seventh of the potential voters or less than half of the residents of the city itself. The ordinary voter, in short, exercised his privilege only when very important and difficult problems were to be discussed and decided and could scarcely have taken off 40 days in the year for regular attendance.

Plato properly summed up the situation in observing that the masses "support themselves by their labour and do not care about politics, owning very little property; this is the largest and most powerful element in a democracy when it is assembled."[19] If we could have looked out over the Pnyx when the assembly was in session we would have seen primarily urban residents with a lesser sprinkling of rural voters, but both elements were of some substance, though we may accept Plato's qualification that occasionally the well-to-do would be outnumbered, as is suggested in some but not all of Demosthenes' orations.[20]

No one, interestingly enough, seems ever to have raised the question as to how a man could prove his right to vote in the assembly. The answer must be that, if challenged, he would appeal to neighbors, friends, and relatives; ultimate attestation could be presented only by means of the deme registers of citizens, which were hardly available except after considerable delay. Very probably, also, the faces would soon have become familiar if one were a regular attendant, as is true in modern democratic gatherings in the academic and civil worlds, where only a minority make the effort to participate in public business.

Supplementary Note

Only with reluctance have I accepted in the text above Finley's figure of 60,000 slaves for Athens just before the Peloponnesian War broke out, even though this is the lowest number which is currently advanced (Gomme gave 115,000, for example).[21] The one relevant, contemporary statement is Thucydides' report that over 20,000 slaves fled to the Spartan post of Decelea in northern Attica in the last stages of the war, but this is given far more weight than it deserves; did the Spartans keep such accurate body counts, especially of slaves?[22] Once again the authority assigned to Thucydides' magisterial approach does not appear justified.

There is, on the other hand, a good deal of evidence to warrant caution in assessing the number of Athenian slaves. In a careful exploration of the social structure of Athens in the later fourth century Jones concluded that at that time there was a maximum of 20,000 slaves as against 21,000 male citizens in the census carried out by Demetrius of Phalerum (317–07) and 10,000 metics.[23] Earlier in the century the Eleusinian building accounts give 20 citizens, 54 metics, 20 slaves; just before the end of the fifth century Erechtheum records attest payment to 20 citizens, 35 metics, and 14 slaves.[24] Even if one assumes slaves were by and large less well trained in stone-cutting, the proportions are greatly at variance with the usual ratios given for metics and slaves as against citizens.

One might argue that the bulk of slavery lay in household and rural service, but this too needs thoughtful assessment. As Ferguson succinctly but wisely put it, "The majority of the farmers had to till the land with their own hands. Many citizens—at least one-third of the whole, in all probability—had to earn their living by selling their labor," and it should be kept in mind that *thetes* were unlikely to have the capital to buy slaves or to be able to feed them. In the United States nearly three-quarters of the white residents in the southern states before the Civil War had no direct ties to slavery.[25] Even the well-to-do in Athens were not wealthy by Lydian or Persian standards; the largest estates known in Athens were in the range of 30 hectares.[26] Agricultural slavery, again, has often been unwarrantably exaggerated; as Aristotle noted, "the poor man, not having slaves, is compelled to use his wife and children," and rural plots were not often large enough to support a significant number of slaves.[27] I would be happier with an estimate that Athens in the fifth century had no more than some 30,000 slaves as a maximum figure, not a minimum, but then my view that the place of slavery in classical societies has been much overstressed is generally rejected.[28]

IV

Functions of the Assembly

The major areas of assembly activity included cult, finances, elections, military and naval preparedness, foreign policy, and justice (though only in political cases as a rule). The vehicle by which decisions were reached was usually the passage of legislation, which need not be considered in itself but rather in each sphere of action. In obedience to Aristotle's dictum that the first care of a state is for its gods, we may properly begin with cult even though the Athenians were not continuously and significantly committed to debate in this respect.[1]

Religion

Oaths taken at Athens to solemnize treaties and other public business invoked Zeus, Apollo, and Demeter, but the divine protectress of the Athenians was Athena, the Virgin goddess worshipped on the Acropolis from early times. Sacrifices in the days of Pericles took place at the great altar in front of the Parthenon, though Athena's cult statue was in the Erechtheum. Each year the citizens celebrated the Panathenaic festival as the high point in the religious calendar; every four years an enhanced Great Panathenaia occurred.

The first item of business on the agenda of the second meeting of the assembly in each prytany took up religious matters (*ta hiera*), but it does not appear that such issues normally excited the citizens deeply; the exception in the fifth century was the witch-hunt in 415 for the criminals who mutilated the images of Hermes which stood everywhere and also allegedly profaned the Eleusinian mysteries in drunken revelry. The vehement reaction of the Athenians to these charges, which involved Alcibiades among others, was in part the result of tensions rising from the launching of the great expedition to Sicily; Alcibiades himself was recalled from Sicily to stand trial but fled to Sparta. The unfortunate culprits who were caught were condemned in the law courts to execution or at least the confiscation of their property, even to the sale of objects for obols.[2] On the whole, however, most sacrifices and cult observances were safely in the hands of the archons, inherited priesthoods, and local neighborhood groups, the fundamental matrix of Athenian society, and the ordinary citizen was not expected or required to attend religious ceremonies or even to express reverence. There was, indeed, a marked decline in popular attention to Athena across the fifth century.[3]

The assembly, nonetheless, had ultimate authority and responsibility in all matters concerning the public religion of state, including the building of temples, institution of new cults such as that of Asclepius during the Peloponnesian War, sacrifices, festivals, and other aspects which were often considered primarily in their financial light.[4] Boards for the festivals were elected or chosen by lot; after the processions and competitions connected with the annual celebration of Dionysus the assembly met the very next day to vote whether all had gone properly and assessed fines or wreaths. Citizens were ready to involve themselves in the most minute detail, as in a decree approving the design for the door of the new temple of Athena Nike;[5] when a fresh set of instructions for the offering of grain at the Eleusinian festival was drawn up about 422 the assembly saw fit to amend them at several minor points during the debate on ratification.[6] Finally the assembly exercised the power to tap funds from the temple treasuries during emergencies in the Peloponnesian War.

Finance

Financial and economic issues, which command a great deal of attention in modern legislatures, were not a great concern for the Athenian assembly though political leaders could be expected to know something about revenues and expenditures.[7] Greek states did not intervene lightly in the economic life of their communities save at least at Athens with respect to overseas supplies of grain, vital for its citizens, and a matter considered on the agenda of the first meeting of the assembly in each prytany. This became a more difficult problem in the fourth than in the fifth century; only by the later date could it be said that "the foreign policy of the Athenians was largely a grain policy."[8]

Religious expenditures, as already noted, could be approved by the voters; otherwise the assembly appears in the epigraphic record for the fifth century primarily in relation to the administration of the empire—these matters, after all, were often worthy of public record on stone. Thus the people decided to devote the surplus revenues stored on the Acropolis for its embellishment, and thereafter a sweeping ordinance directed that all subject states use standard weights and measures and Athenian coinage.[9] Again the major revision of the Athenian tribute lists in 425, which much increased the burden of empire, was ratified in a public decree, perhaps "the strongest decree that has survived from the fifth century."[10]

The collection of the tribute, however, was more a function of the council, which employed overseers and traveling inspectors, backed if necessary by an unpleasant visitation of an Athenian squadron to a delinquent dues-payer. Financial affairs as a whole were directed entirely by boards chosen by lot, which were also supervised by the council. Athenians were far from trusting each other with public monies; auditors checked every magistrate who had dealt with state funds at the end of his term, and every prytany a committee of the council also examined public accounts. Even so, Aristotle ruefully observed that men in his day wanted to hold public office continuously, "moved by the profits to be derived from office and the handling of public property." By the fourth century generals and orators could also make considerable advantage from their posts, probably from foreign bribes as was alleged earlier for Themistocles.[11]

One last example of interference in the economic sphere, this time concerning Athenian markets, lies outside our period; but the decree passed in 375/4 also deserves note in its very specific requirements that genuine Athenian coinage be accepted and that public officials test such coins and deal with counterfeits. Legislation in this detail on economic problems is not yet attested for the fifth century, yet there is no reason to assume that the assembly could not have taken action, especially concerning coinage, a state responsibility regulated elsewhere in fourth-century decrees.[12]

Public Officials

One of the unique characteristics of Athenian democracy was its heavy reliance on choice by lot of public officials from the archons down, for only one year and usually in boards of 10 assigned very specific duties. The urban commissioners, for example, had a body of slaves who each morning removed those who had died overnight in the streets. The Athenians pushed democracy to an extreme rarely seen in later times, but they were careful in practice; all officials were examined before entering office, were checked by the council as they performed their tasks, and could not lay down their posts at the end of the year until inspected and approved by the council.

No citizen in his right mind, however, would entrust his life in battle to officials so chosen, so military leaders were always elected by vote and could serve as many years as the assembly accepted them: 10 generals (who also acted as admirals and very commonly were political leaders of the assembly), two commanders of the cavalry, 10 leaders of the tribal contingents, 10 squadron officers. Modern students, accustomed to party and factional debate in elections and the conduct of public business, may well feel that similar patterns must have occurred at Athens, but this is a problem best left for consideration in the next chapter.

The assembly also deemed it wise to elect by vote annually specialists such as the city architect, the superintendent of the water supply, and the board of naval architects; supervisors of the dockyards could more safely be chosen by lot. Direct responsibility for maintaining the naval establishment, on which Athenian external power rested, was in the hands of the council, which oversaw among other matters the building of new triremes as ordered by the assembly.[13] Ship captains (*trierarchs*) were appointed for one year, probably by the generals, from the wealthier citizens.[14] The financial burden of the navy was more easily met in the fifth century, when Athens could draw on imperial revenues, than in the fourth, but even then the state still contrived to maintain the largest Greek fleet.

Down to the last decades of the fifth century, as just noted, the leading general along with his colleagues was the most influential figure in the assembly and was from aristocratic stock; despite this weight he could be checked by the process of *euthyna* as well as the mighty machine of *eisangelia*. Only with Cleon in the Peloponnesian War did a separation of the two functions begin to be apparent, though Cleon served several times as general; in the fourth century generals and public orators were normally distinct.

Generals had direct access at least in wartime to the council and the assembly; epigraphic evidence attests that they could propose decrees.[15] Pericles once prevented the assembly from convening when the citizens were angered by the Spartan devastation of the countryside early in the Peloponnesian War,[16] and Cleon was reported to have come tardily and adjourned a meeting on his own initiative. As Aristophanes ironically describes his power at the height of his popularity, "Of all these people you shall personally be lord and master, and also of the marketplace, the harbors, and the Pnyx. You will tread the Council under foot; you will clip the generals; you will imprison people, arrest them, taste sex in the Prytaneion."[17]

Commanders sent abroad, however, were often restricted very specifically by the dictates of the assembly; the small expedition despatched to Corcyra just before the Peloponnesian War was ordered not to engage with the Corinthian fleet save if Corcyra itself were attacked, so as to prevent a breach of the treaty of peace with Sparta.[18] True, the Athenians promptly violated their orders when a naval battle found the Corcyreans hard pressed. At other times generals were less limited; those in command of the expedition to Sicily had power "to act as they thought best about the size of the force and the whole expedition."[19] Although one general such as Pericles might have predominant influence, it is far from certain that the generals always agreed unanimously; certainly in the ill-fated attack on Syracuse Nicias, Alcibiades, and Lamachus had very different views on the proper course of action. Much later Demosthenes was to complain that only one general in his day was actually sent in the field: "you are like makers of terracotta statuettes: you make regimental commanders and squadron leaders for the market, not for war."[20] But one must always keep in mind that the assembly had ultimate powers on making war and peace, as was specified in the codifying decree of 403, and also had to vote for expeditions by land and by sea.[21]

Foreign Policy

Expeditions, treaties, decisions on war and peace—these are the stuff of which foreign policy is made, and it was in this area that the assembly showed itself as master all across the fifth century. Debate on granting aid to the Ionian rebels was ended by a vote in the assembly, as Herodotus reports. After Themistocles had persuaded the assembly to devote the unexpected funds from the silver strike at Laurium to the creation of a large navy, the Athenians stood firmly behind him in the strategic decisions as to how to meet the Persian invasion even to the extent of abandoning ancestral shrines and graves rather than submit to Xerxes. Thereafter the assembly in surviving decrees bulks large in the establishment and consolidation of the Athenian empire.

Especially in the pages of Thucydides it is the assembly which always appears as taking the major steps. It debated an alliance with Corcyra and sent out ships to help this state; it passed the decree which banned Megarians from the harbors and marketplaces of the empire, a direct challenge to the allies of the Spartans; it meditated on the proper punishment of the rebels at Mitylene early in the Peloponnesian War. One could go on thus across the course of the war and always find the assembly in the foreground. Sometimes the voters were of two minds and had to meet more than one day to settle on a course of action; in the case of the Mityleneans they were thus persuaded at a second meeting to reverse their brutal decree to execute all the rebels.

It almost appears as if the council had no role, for Thucydides only once mentions it in connection with Alcibiades' scheme to hoodwink Spartan envoys into telling one thing to the council and the next day something else to the assembly.[22] But proper constitutional procedure must have been regularly followed, even if the council only issued an open *probouleuma* on a topic, leaving decision to debate in the assembly. Here the discussions could be acrimonious, as in those which led to the preparation of the great expedition to Sicily in 415; it was the assembly which had to vote to send reenforcements on the desperate appeal of Nicias, which only made the final disaster more catastrophic. When war was resumed in the Aegean and the Spartans found naval operations difficult, they offered several times to make peace, but the desperate assembly always refused and so led the Athenians on inevitably to their final collapse.

Justice

Whereas foreign policy was a continuing issue for the citizens meeting in the assembly, the provision of justice was in the hands of the citizens serving in the law courts in juries (*dikasteria*) of 201 or more—always an odd number since a majority vote was normally expected for a verdict. Jurors were volunteers over the age of 30 and so were likely to be an older body than the assembly, composed of voters of 18 years or more; but the common assumption that the jurors were more conservative is not very well supported by the evidence on trials in the fourth century. Inasmuch as almost all public decisions by magistrates and the assembly could, even so, be appealed to the courts, they may be called in Ferguson's words a "brake" on rash steps.[23]

Both the council and the assembly could refer cases to the courts.[24] In the sixth prytany it was ordained that the assembly could consider three charges for "malicious prosecution" (*sycophancy*); if so voted by the people the charges were turned over to the jurors.[25] The assembly itself, however, could hear and judge other charges, as when Phidias was accused of embezzling gold and ivory intended for the statue of Athena in the Parthenon;[26] and apparently a decree of the people offered a reward for the apprehension of Diagoras of Melos, on the grounds that he maligned the Eleusinian mysteries in 415/4.[27]

Most important were "denunciations to the people" (*eisangeliai tou demou*) on three or more accusations: treason, attempt to overthrow the democracy, and corruption, all of which were really vehicles for partisan attacks on unpopular figures or leaders of the democracy.[28] Here again, if the people accepted the denunciation it was usually tried in the law courts, but the assembly could, if so persuaded, turn itself into a forum for judicial decision and could assess the death penalty if a culprit was convicted.[29]

This responsibility was transferred from the Areopagus to the council of 500 and the assembly by Ephialtes according to a current theory, as noted in the second chapter; but the assembly had certainly felt able 25 years earlier to hear and pass judgment on charges raised against Miltiades. As a general at Marathon he had persuaded the titular commander, Callimachus, to march down into the plain and attack the Persians, who were routed and fled to their ships. Militiades, flushed with glory, then promised to take the island of Paros with all its purported riches and was granted authority by the assembly to lead an expedition of 70 ships. The attack failed, and Miltiades was seriously injured in a thigh. On his return to Athens his opponents "impeached him before the people, and brought him to trial for his life, on the charge of having dealt deceitfully with the Athenians." Beneath the surface there may also have lain the Athenian fear of tyranny, for Miltiades is said to have been called *tyrannus*. By the time of the trial he had to be carried before the assembly, by reason of gangrene, and could not speak in his own defense; the assembly was satisfied by levying the heavy fine of 50 talents, which his son Cimon paid after his death.[30]

Thereafter the assembly was generally content, save in one or two ill-reported cases, to refer charges of *eisangelia* to the courts down to the most remarkable example of popular justice or perhaps more accurately lynch law. This occurred late in the Peloponnesian War, in 406 when the temper of the Athenian people had been hardened by years of adversity.

A few years earlier, to set the stage for the events of 406, the Spartans had entered into a devil's bargain with the Persian satraps of Asia Minor to turn over the Greek cities on its west coast in return for funds to build and man a fleet, this despite the fact that the battle cry of Sparta had been its effort to liberate the Athenian subjects from their "slavery" (*douleia*). The first Spartan effort resulted in a disaster at Cyzicus in 410; the Spartan despatch home was a laconic masterpiece: "Ships gone; Mindarus dead; the men starving; at our wits' end what to do."[31] The Spartans offered to make peace on the basis of the *status quo,* but the Athenians refused; so the Persian satraps had to dig more deeply in their purses to help the Spartans build a new fleet.

This time they were successful, in 406, in bottling up the Athenian battle fleet at Chios. To relieve the blockade the Athenians launched every hulk in the shipyards; crews were provided by slaves (the only time in Athenian history) and the well-to-do hoplites, who also never served regularly as rowers.[32] This scratch flotilla rowed across the Aegean, met, and defeated the Spartans in the battle of the Arginusae islands; but a storm prevented the generals from rescuing those cast into the sea. When this news came home the Athenians were enraged, not over the loss of the slaves but of the hoplites who had manned the oars—if evidence were needed for the general composition of the assembly, as suggested in the third chapter, this reaction should prove which elements usually dominated meetings on the Pnyx.

Only six of the eight generals present at the battle ventured to come back to Athens; they were promptly imprisoned by the council and brought before the vengeful assembly. The debate went on so long that darkness fell before there could be a vote; the council was instructed to draw up a *probouleuma* for the meeting the next day. An enemy of the generals, Callixenus, secured passage in the council of a resolution that the assembly should simply vote to convict or acquit without any further discussion. Socrates, one of the prytanies at the time, refused to support the proposal, but when one citizen objected strongly he was shouted down: "It is shocking not to let the people do whatever they wish." The generals were convicted and promptly executed (one was the son of Pericles); never before nor thereafter was popular sovereignty pushed to such a ruthless conclusion.[33] As a rule the assembly conducted public business in all its varied functions with a proper amount of caution and good judgment.

V

Meetings of the Assembly

By 339 Philip of Macedon had consolidated his power in northern Greece, first by humbling the barbarians on his frontiers, then by taking and destroying Olynthus in 348 despite tardy Athenian aid to the city, and thereafter by becoming master of Thessaly and Phocis.[1] Under the leadership of Demosthenes the Athenians had finally recognized his threat but had been forced to agree to a peace in 346 which accepted Philip's gains.

Now he was ready to advance into central and southern Greece and late in 339 seized the vital outpost of Elatea on the northern fringe of Boeotia. This dangerous move was reported swiftly to Athens; the speedy reaction of its machinery of government in time of crisis is more graphically described by Demosthenes than any similar event in the fifth century, though probably much the same sequence took place then in response to an emergency:

> It was evening when a messenger came to the prytanies with the news that Elatea was captured. They immediately got up in the middle of dinner and expelled the occupants of the stalls in the market place and burnt the hurdles, while others sent for the generals and summoned the trumpeter; and the city was full of tumult. The next day, at dawn, the prytanies summoned the council to the council chamber, and you went to the assembly, and before the council had opened proceedings and voted a reso-

49

lution the whole people was sitting on the Pnyx. And then the council came in and the prytanies announced the news that had been brought to them, and introduced the messenger, and he spoke. Then the herald asked: "Who wishes to speak?"[2]

In the fifth century he would have qualified this invitation to begin proceedings by specifying that priority was given to those over 50 years, senior citizens in an era when the average man died in his thirties.[3]

To conclude the story, it was Demosthenes, *prostates tou demou,* who donned the wreath which formally gave him the floor and convinced the assembly to seek alliance with its inveterate enemy Thebes by offering favorable terms and mobilizing the Athenian army as a pledge of sincerity. He then traveled to Thebes and amazingly persuaded the assembly there not to accept the blandishments of Philip but to join forces with Athens. The upshot was a battle between the Athenian and Theban levies, led by a host of generals, and the veteran army commanded by Philip alone at Chaeronea on August 2, 338. Philip won the hard-fought battle, which essentially ended the independence of the Greek state system. Toward Thebes he was harsh, toward Athens much more lenient; he might be able to use its navy and admired its great cultural tradition. The assembly accepted the alliance which he proposed and awarded citizenship to Philip and his son Alexander, but it did not turn on Demosthenes, who was given the great honor, albeit a perhaps unpleasant one, of delivering the state oration in praise of the dead of Chaeronea.

Procedures of the Assembly

By the time the council entered the assembly in Demosthenes' account, preliminary purification of the Pnyx would already have been carried out as well as lustral prayers and curses on the Medes, abettors of tyrants, and other evil doers. Usually if a weasel ran across the hillside a meeting would have been postponed, but such an unfortunate omen presumably did not occur on the day of the session to decide Athenian reaction to Philip's threat.

When the debate proper began, it was supervised by the chairman of the assembly (*epistates*), a man chosen by lot from the prytanies of the period to be chairman of the council also. As has been observed, the Athenians here too preferred democracy to efficiency; rather than having a chairman elected for the purpose who might unduly influence the assembly they were "prepared to risk having one who could not distinguish between an amendment and a substantive motion."[4] True, his fellow prytanies sat immediately below him and by nods or other motions might help guide him in presiding over the assembly of several thousand citizens; in decrees the *epistates* is named immediately after the secretary, who was elected, not chosen by lot, so as to ensure that someone was named who could read out messages and *probouleumata* and also record in writing decisions of the assembly, duties which not all citizens would have been able to perform. Ostracisms and elections, however, were conducted by the archons.

The agenda for a normal meeting was drawn up by the council and published four days earlier. Many of the items to be considered were laid down by the standing rules of procedure; others were properly introduced by a preliminary *probouleuma* of the council. A good deal of the assembly's business probably passed without any objection, but as suggested in Thucydides' reports of several important meetings before and during the Peloponnesian War fierce debate might ensue.

Here too there were rules which are summarized in a speech by Aeschines:

> Speakers in the council or assembly must keep to the subject, must treat each subject separately, must not speak twice on the same subject at the same meeting; must avoid invective, must not interrupt another speaker, must not speak except from the *bema,* must not assault the *epistates.* For each offence the *proedri* may impose a fine of up to 50 drachmae, or for a greater penalty they may refer the matter to the next meeting of the council or assembly.[5]

Even so, a speaker needed a strong voice and commanding presence to make himself heard. In Plato's *Protagoras* Socrates gives a lively picture of how sessions could proceed:

I, like the other Greeks, think that the Athenians are wise. Well,
I see that when we gather for the assembly, when the city has to
do something about buildings, they call for the builders as advisers and when it is about ship construction, the shipwrights,
and so with everything else that can be taught and learned. And
if anyone else tries to advise them, whom they do not think an
expert, even if he be quite a gentleman, rich and aristocratic, they
none of them refuse to listen, but jeer and boo until either the
speaker himself is shouted down and gives up, or the sergeants
at arms, on the order of the prytanies, drag him off or remove
him. That is how they behave on technical questions. But when
the debate is on the general government of the city, anyone gets
up and advises them, whether he be a carpenter or a smith or a
leather worker, a merchant, or a sea-captain, rich or poor, noble
or humble, and no one blames them like the others for trying to
give advice.[6]

As Socrates notes, debate in the assembly was not always or-
derly, and would-be speakers sometimes had to be ejected by
force, either by the prytanies or by the Scythian archer police,
who also attended meetings of the council to keep proceedings
there under control. Difficult issues might not be settled in one
meeting and had to be carried over to the next day, or on occasion
the assembly could be convened again to reverse a previous decision, as in the question of the punishment of the Mitylenean
rebels.[7] In the passage just quoted it also is to be noted that essentially any citizen could speak if he could gain the floor; *isegoria*
was "considered by the Athenians to be a cornerstone of democracy and so it was," even though proceedings in the assembly certainly could not allow everyone to give his views.[8]

Eventually matters came to a vote, which was settled by show of
hands;[9] if the *epistates* was uncertain where the majority lay he
could call on his fellow prytanies to help in the assessment. So the
number of votes on either side could not have been reported in our
sources unlike the outcome of trials where the jurors had ballots.[10]
The system does not seem to have caused uncertainties; Aristotle
considered the Spartan method of voting for ephors by shouting
childish.[11] Thereafter came as a rule a decree setting down formally the decision of the citizens. This was drawn up by the secre-

tary and sometimes, fortunately for modern studies of Athenian democracy at work, published on stone.[12] Each decree had to have the name of the citizen who proposed it, either on his own volition or acting for the council so that if later reflection judged it unwise he could be charged in the courts with a *graphe paranomon.* Analysis of surviving decrees shows that often they directly reproduced the council's *probouleuma,* but others were revised on the floor or even replaced by a different resolution. The citizens assembled on the Pnyx had the final voice.

How did they gain the information which they needed to judge, say, the advisability of a treaty with Sitalces of Thace or to assess the relative merits of staying on good terms with Corinth as against aiding Corcyra before the outbreak of the Peloponnesian War? The easy answer is that they hearkened to their leader of the day and voted as he recommended, but while this may have been true to a considerable extent when Pericles was dominant, it is not a convincing explanation at other points. Cleon and Diodotus gave opposite advice on how to deal with the Mityleneans; Nicias and Alcibiades diametrically disagreed with each other on the desirability of the Sicilian expedition; much earlier Aristides and Themistocles opposed each other. True, Thucydides laid the blame for the Athenian defeat in the Peloponnesian War on the blunders of the citizens, following poor leaders who flattered them to gain ascendancy,[13] but there was a multitude of financial, military, and other problems, including the disaffection of the Athenian subjects, which must be put on the scales to explain the final collapse—the quality of leadership was not the only factor.

In the end one comes back to the fact that the citizens made the decisions, and unlike modern representative democracies did so directly by their votes in the assembly. The issues they faced, to be sure, were much simpler than those in the modern world, and in the marketplace they could gain information and misinformation on which to base their judgments; in 415 the Athenians were pictured as "sitting in the wrestling grounds and public places, drawing on the ground the figure of the island [Sicily] and the situation of Libya and Carthage."[14] As in voting the Sicilian expedition, the assembly could make dire mistakes and was often misled by men seeking public power, a charge made by Aristophanes also, but even in his plays it was "amenable to reason and capable of correction and improvement."[15]

The Abortive Revolution of 411

Certainly the Athenians were deeply attached to their democratic principles and were loath to yield them. In the closing, grim years of the Peloponnesian War, however, they were forced to do so twice, in 411 and again in 404.

By 411 Athens was in dire straits. Abroad its generals had to face a Spartan fleet much enlarged by Persian funds and yet had the mission of keeping down the subjects, who had lost citizens at Syracuse and were outraged by the imposition of a 5% tax on all imports and exports in the empire; only Samos was to remain loyal to Athens down to the bitter end of the war. At home the war-weariness of the citizenry is visible in the tone of the comedies of Aristophanes and the tragedies of Euripides. All elements were suffering psychologically and economically, but the group which was hardest hit was that of the rural landlords, the hoplites, who were barred from their farms by the standing Spartan garrison at Decelea in northern Attica. If an opportunity arose there were conservatives willing to try to seize control and bring an end to the war.

The man who instigated a plot was Alcibiades, who had the ear of the Persian satrap Tissaphernes at Sardis. In 415 Alcibiades had been recalled from the expedition to Syracuse to stand trial for his purported role in the profanation of the Eleusinian mysteries; fearing for his life he had escaped to Sparta, where he gave valuable counsel against the Athenians. But alas he seduced the wife of a Spartan king and had to flee again, this time to the Persians. Now in 411 he sent word to the Athenian fleet, lying for the moment at Samos, that he could persuade the Persians to give aid to Athens if it would "abolish the villainous democracy which had driven him out" and also to provide pay for the sailors.[16] The generals and other officers were won over; "the multitude were at first dissatisfied with the scheme, but the prospect of the King's pay was so grateful to them that they offered no opposition."[17] Envoys were sent from Samos to Athens, led by Peisander, who urged the assembly to accede to the Persian wishes. "Do not let us be dwelling on the form of the constitution, which we may hereafter change as we please, when the very existence of Athens is at stake."[18]

The citizen body was thrown into confusion and dismay at the murder of stalwart democratic leaders, conducted by the oligarchic clubs, and by the swirling rumors; no one could know who was in the plots. The uncertainty both at Athens and at Samos need not be described at length; the end result was the ability of the true oligarchic leaders at Athens, Peisander, Antiphon, and others, to move openly. Accordingly, they summoned the assembly to meet, not on the Pnyx but a mile outside the walls at Colonus; since there was a Spartan garrison nearby at Decelea only the hoplites dared attend and duly voted to entrust the state to a group of five men, who were to appoint a band of 400. These in turn were to draw up a list of 5000 voters, though this step was never carried out. "An easy thing it certainly was not, 100 years after the fall of the tyrants, to destroy the liberties of the Athenians, who not only were a free, but during more than half of this time had been an imperial people."[19] Temporarily, however, the attack on democracy was completely successful. The Four Hundred even burst into a meeting of the council and peremptorily dismissed it, but punctiliously paid the wages of the councillors for the rest of the year. They themselves occupied the council chamber and proceeded to govern Athens, opening negotiations with Sparta.

The reconstruction of the government soon began to fall apart. At Samos two generals together with the trierarch Thrasybulus and a common sailor, Thrasyllus, restored democracy and executed the major supporters there of oligarchy. Developments at Athens and in the Piraeus were protracted and complicated; the result was that the conservatives fell to feuding over who could gain the advantage. The moderates, led by Theramenes, picked up courage and began to hold assemblies on the Pnyx, which deposed the Four Hundred, "re-appointed supervisors of the law, and by a series of decrees established a constitution." Thucydides goes on to give his judgment, "This government during its early days was the best which the Athenians ever enjoyed within my memory. Oligarchy and Democracy were duly attempered."[20] In 410 full democracy was restored.

Alcibiades, incidentally, succeeded in getting his recall to Athens and was even elected general, but in 406, when one of his subordinates lost an engagement, he again went into exile and never came home again. The portrait which Plutarch drew of him in connection with his stay at Sparta is one of the most remarkable sketches of the biographer's pen: he could be a chameleon and adapt himself to all types of environment even though his career was one of wasted opportunities.[21]

Subversion of Democracy in 404

Events in 404 moved in a more ruthless fashion. After Athens had surrendered, Lysander, the Spartan admiral, himself attended a meeting of the assembly and ordered it to approve a proposal that 30 men be appointed to draft the "ancestral laws" and govern the state.[22] Toward the end of the fifth century there had been a considerable interest in "restoring" a more conservative form of government, the *patrios politeia* embodied in the laws of Draco and Solon, a movement on which Critias and his coadjutors eagerly seized.[23] The democratic organs were abolished, even including the law courts, and the reforms of Ephialtes which had clipped the powers of the council of the Areopagus were rescinded. Using 300 club-bearers to enforce their directives the "30 tyrants," as they soon were named, ruled Athens despotically. Five thousand citizens were banished, 1500 executed—the moderate leader Theramenes was torn from an altar in the council chamber and dragged across the Agora to drink his fatal cup of hemlock. To support their control the 30 secured a Spartan garrison on the Acropolis.

These excesses quickly led to a democratic reaction. Early in 403 Thrasybulus, the erstwhile trierarch active at Samos in 411, and a band of 70 followers seized the frontier fortress of Phyle; sufficient numbers of supporters joined him so that he could move down into the plain and occupy the Piraeus. Critias led the conservative forces in an attack which failed; Critias himself, the most vengeful of the 30, was killed. As the power of the oligarchs weakened they appealed to Lysander, who returned to Attica and blockaded the Piraeus. Yet a vital shift in the backgound was occurring; Lysander's influence at Sparta was waning, and the Spartan king

Pausanias superseded him in Attica, reversing the policy of up-holding an unpopular tyranny. The remainder of the 30, with their adherents, withdrew to Eleusis; the democrats reoccupied Athens, and the Spartan garrison yielded the Acropolis. During the year 403/2 full democracy again came into force; in a decree moved by Teisamenus it was ordered, "Let the Athenians be governed by their national customs."[24] The restoration was not challenged again for almost a century.

In 401 the oligarchic occupation of Eleusis was ended, and Attica was reunited. The oath of reconciliation then sworn "consisted of a simple asseveration, 'We will remember past offenses no more'; and to this day the two parties live amicably together as good citizens, and the democracy is steadfast to its oaths."[25] As has recently been observed, this is "one of the most inspiring episodes in Athenian history, if not even in human history" in view of the moderation with which the oligarchs were treated.[26]

Recently a powerful argument has been advanced that a fundamental change in the nature of democracy and the role of the assembly had actually taken place, to wit, that instead of a structure of popular sovereignty Athenian government was now checked by the rule of law as codified in the major revisions of 403/2; thenceforth "no decree of the people shall supersede a law."[27] While it is true that laws (*nomoi*) were now more carefully distinguished from decrees (*psephismata*) and were drawn up by a special board of *nomothetai,* their formal adoption required a vote of the assembly, and on the whole the summation in *Constitution of the Athenians* remains true that "the people have made themselves masters of everything and administer everything through decrees of the Assembly and decisions of the law courts, in which they hold the power. For even the juridical functions of the Council have passed into the hands of the people."[28]

Why Did the Assembly Succeed?

The problem involved in this query does not require an evaluation of Athenian democracy as a whole, a subject which has been explored in many works, or the theoretical defense of democracy. For the latter, indeed, there is very little evidence in the fifth century or later ancient sources; from the anonymous conservative called the Old Oligarch or Pseudo-Xenophon, who wrote a brief, bitter tract, on through Plato and Aristotle, Athenian democracy was severely criticized, as it was to be later by Cato the Censor and most modern political analysts into the nineteenth century. It is not true, however, that the principle of democracy was always attacked;[29] apart from the famous Funeral Oration Thucydides also briefly reported a defense by the Syracusan leader Hermocrates, and in Herodotus there is an interesting view expressed by one of Darius' fellow conspirators. Otanes does not directly describe Athenian democracy, but the composer of his defense of democracy (probably Herodotus himself) does cite one of its cardinal principles, that of holding officials to account for their actions (*euthyna*).[30]

The assembly, after all, was only one part of the Athenian system of government, which embraced the board of 10 generals and other officials, the council of 500, and the law courts; the latter were in the opinion of Aristotle the most important element in Athenian democracy.[31] The one issue which concerns us here is how a mass of citizens numbering in the thousands, seated on the rocky hillside of the Pnyx, could operate efficiently. Often students have expressed amazement or even doubt that it could make rational decisions without the guidance of outside factors, but the Danish scholar Hansen has drawn an instructive parallel with the mass meetings in several Swiss cantons which transact public business expeditiously and usually wisely.[32] So we may conclude that the Athenians could possibly have done as well, but first it may be useful to consider some of those external elements which have been adduced in efforts to explain how the assembly did function.

A dominant role thus has often been assigned to the "leaders of the people" (*demagogues*—not as pejorative a term in ancient Athens as in modern contexts). As was noted earlier, the assembly attended carefully to their advice, and it is difficult to visualize how it could have operated efficiently without that advice. Yet on the one hand political figures could not always be entirely frank and had to couch their views with a certain degree of subtlety to gain assent; and on the other there was always a tension, conscious or not, between the existence of these leaders and the demands of the community for general cohesion.[33]

Yet, though Athens was a democracy in which even the *thetes* could occasionally exercise considerable influence, the men who spoke with greatest authority were usually aristocratic citizens from the time of Solon to Demosthenes. Technically this statement can be faulted in one respect: Athenian politicians did not have inherited titles, nor were they even distinguished in dress as were Roman *nobiles* in their purple-edged togas. Still, as men who commanded respect, descended from families of good repute, they were indeed essentially aristocrats. In the later fifth century Cleon, Cleophon, and others were of a somewhat different stamp, but the charges against them of low birth and slurs as to their professions of tanner or lye-maker are in part examples of the innuendo which afflicted Athenian political debate;[34] equally scurrilous remarks were hurled by Demosthenes and his peers at each other.

The Athenians, even so, kept their leaders on a tight rein and could easily become suspicious of tendencies to arrogance, as in the well-known case of Themistocles. There were undoubtedly more charges of *eisangelia* than those of which we happen to hear, and almost every *prostates tou demou* eventually suffered ostracism down to its practical abolition, save for Pericles. In the end, moreover, it was the citizens who decided which advice to follow and approve decrees accordingly. So we must look further to determine how they played their role in public affairs.

Some years ago Hopper, in an inaugural lecture on the basis of Athenian democracy, suggested that "a man who knew his deme business would not be lost in state business," an interesting view but one that does not square with the facts.[35] There is sufficient information on the men who were active in deme affairs to demonstrate that they had very little place in the administration of the state, perhaps one archon, no generals, only three or four proposers of decrees out of 157 known examples in the fourth century.[36] The demes, again, which varied from 150 to 1500 members, were not directed in a democratic fashion; the leaders here were the local well-to-do landlords.[37] They might well have attended the assembly on contentious matters, but they had no significant power in its deliberations and votes. "Those active at deme level and at city level were, broadly speaking, different people."[38] Cimon might cultivate his fellow demesmen, but political leaders in Athens itself needed a broader base of support.[39]

The same caution must be applied to any theory that Athenians were split into lasting factions, though historians accustomed to modern political organization might find it appealing. In the make-up of the council, groupings of tribal members have been sought out, but these could not have existed for more than one year inasmuch as membership in the council rotated every year.[40] Even in the assembly men seem to have seated themselves more or less at random; there is no real evidence for arrangement by tribes and trittys.[41] Only two instances to the contrary can be cited for the fifth century. Thucydides son of Melesias, the opponent of Pericles, thus saw to it that his adherents sat together; in the *Ecclesiazusae* Praxagora had her supporters take the first seats, facing the prytanies, but this was an effort to disguise the fact that they were women who had filched their husbands' clothes and donned fake beards.[42]

Clubs of young men existed all across the sixth and fifth centuries, mostly but not altogether composed of aristocrats; metics took part in the parody of the Eleusinian mysteries in 415. The purpose of the clubs, however, was largely to celebrate those symposia depicted on Athenian vases with drinking cups and attendant scantily clad girls; in the context of procedures in the assembly members of these groups could have influenced proceedings mainly by boos and cheers. Only in the upheaval of 411, when law and order were temporarily in abeyance, does one hear of political hooliganism and worse from the clubs.[43]

There is, to be sure, clear testimony that the Athenians were split over the years on two major themes, the rich as against the poor and the rural population as opposed to urban citizens.[44] These divisions were not entirely conterminous, but in general it is true that the well-to-do were largely landlords; Cephalus and other metics might conduct much of the business of Athens and its port but only exceptionally could own even their homes, for Attic land was reserved for citizens. At times decisions on war and peace did depend on whether rural voters, fearing devastation, or the urban *thetes,* desiring the pay of active service, happened to dominate a meeting of the assembly.

A much more convincing case can be made—and often has been—that the council steered the assembly virtually as its master, "an active and indispensable body."[45] The assembly could take up only matters which the council laid before it, and analyses of decrees from the 450s on show clearly that very often the assembly simply approved a *probouleuma* presented to it in final draft form. Undoubtedly there were scores of decrees which did not warrant being published on stone and went through on the nod; these were merely recorded by the secretary in the state archives, carefully preserved from the later fifth century onwards.[46] Yet there were decrees which had been altered on the floor of the assembly, and Thucydides describes the debates on critical problems as taking place in the assembly with almost no mention of the council; indeed the one decree which he cites verbally does not mention the council, a rare visitor to his text.[47]

Of two recent studies on the council one is ambiguous in treating the relations of the council and the assembly;[48] the other assumes that the council was "the chief organ of government insofar as the making of policy, foreign and domestic, was concerned."[49] It is, however, not difficult to find other analyses which disagree sharply with this sweeping judgment: "Athenian policy was in fact made in the assembly, and not just ratified by the people" in obedience to the council.[50] This is in my opinion closer to the truth; the council had to keep an eye on what the assembly would approve, and the two normally conducted public business in harmonious tandem.

In sum, neither leaders nor participation in deme affairs, factionalism, clubs, and the role of the council suffice to explain why the assembly operated efficiently as a rule. We must always come back to look at those thousands of citizens who sat on the Pnyx 40 or more times a year and listened to Themistocles, Cimon, Pericles, and other "leaders of the people," for the voters are the reason why the assembly was successful in reaching decisions on the problems of foreign and domestic policy placed before it. They did not do so without contention and violent disagreement and at times had to meet a second day to alter or reverse earlier votes, but they did face issues of fairly simple character which men of sound judgment and political sensitivity could master. As was suggested in an earlier chapter, attendants at the assembly tended to be a stable base, not easily deceived, and capable of making up their own minds.[51] First Solon and then still more Cleisthenes gave them the powers necessary to govern the state; across succeeding decades they became ever more practised in doing so. Democracies are often said to be prone to follow poor leaders, seeking popularity, and to make unwise decisions as a consequence, both of which are charges that are occasionally levied against the Athenian assembly and also more recent structures of government;[52] but as in modern times so in the fifth century the Athenians generally exercised their powers thoughtfully save at some points in the stresses of the Peloponnesian War. During its last stages they were temporarily deprived of authority, yet in each case the assembly was quickly restored to its previous position, adequate testimony that citizens considered it a successful rudder of the state.

The structure which was thus bequeathed to the fourth century, after the developments from Cleisthenes onward, deserves summation. The assembly was thus the sovereign source of decisions in Athens and could in the end act as its participants deemed best for the state; as we saw in Chapter III, however, less than 20% of the population of Attica could take part, and in practice only a minority actually attended meetings of the assembly—but then in the perfected democracy of the United States in the twentieth century elections especially on the local level summon only a very small percentage of registered voters. The procedures in meetings of the assembly had become well defined and fully accepted by attendants as a rule, though unruly behavior was not something easily checked by rules in any Athenian meeting, both in the theater of Dionysus and on the Pnyx.

Safeguards, however, had been created across the century. Cleisthenes had introduced the system of electing a steering committee, the council of 500, and by a self-denying ordinance the assembly never took up any problem on which the council had not provided it with a *probouleuma,* even though this could be entirely open. The participants in debate enjoyed *isegoria,* one of the vital hallmarks of democratic rights which Herodotus stressed above all else in his account of the Athenian successes in the late sixth century. A passing comment in Plato suggests that commoners who had some useful knowledge on a technical subject were accorded a voice, but that anyone could speak as he wished is doubtful; rights may exist but they are not always exercised, and inadequate speakers could be swiftly shouted down. The assembly listened most easily to veteran advisers, such as Pericles, though as noted in the second chapter Athenian citizens could be ruthless in their treatment of erstwhile masters of the floor.

Another protection against rash action was the introduction of the *graphe paranomon,* writ of unconstitutionality, which could be hurled at a decree and its proposer, when citizens came later to feel the measure had been ill advised. And finally a formal distinction between *nomoi,* lasting regulations, and *psephismata* or decrees had been accepted toward the end of the fifth century, though the degree to which this limited the scope of activity of the assembly must not be exaggerated. To repeat, the assembly remained in the fifth century the focus of political debate and the source of ultimate decisions.

"We do not copy our neighbors, but are an example to them," to quote the Funeral Oration of Pericles as reported by Thucydides. "It is true that we are called a democracy, for the administration is in the hands of the many and not of the few. But while the law secures equal justice to all alike in their private disputes, the claim of excellence is also recognized; and when a citizen is in any way distinguished, he is preferred to the public service, not as a matter of privilege, but as the reward of merit. Neither is poverty a bar, but a man may benefit his country whatever be the obscurity of his condition. . . . We are prevented from doing wrong by respect for authority and for the laws, having an especial regard to those which are ordained for the protection of the injured as well as to those unwritten laws which bring upon the transgressor of them the reprobation of the general sentiment. . . . An Athenian citizen does not neglect the state because he takes care of his own household; and even those of us who are engaged in business have a very fair idea of politics. We alone regard a man who takes no interest in public affairs, not as a harmless, but as a useless character; and if few of us are originators, we are all sound judges of a policy."[53] The degree to which Athenian citizens as a whole shouldered their responsibilities may be exaggerated, but the general spirit of the stalwart attendants at the Pnyx is handsomely defined.

Notes

Introduction

1. W. S. Ferguson, *Greek Imperialism* (Boston, 1913), p. 51.
2. G. F. Schömann, *De Comitiis Atheniensium* (Greifswald, 1819).
3. M. H. Hansen, *The Athenian Assembly in the Age of Demosthenes* (Oxford, 1987); his earlier work, *The Athenian Ecclesia* (Copenhagen, 1983), assembles a number of detailed essays, also on the fourth century.
4. M. Ostwald, *From Popular Sovereignty to the Sovereignty of Law: Law, Society, and Politics in Fifth-Century Athens* (Los Angeles, 1987), p. 524. Among recent general surveys of Athenian democracy one may cite the idiosyncratic study by R. Sealey, *The Athenian Republic: Democracy or the Rule of Law?* (University Park, 1987), and the detailed description by J. Bleicken, *Die athenische Demokratie* (Paderbörn, 1985); there will be more.

Chapter I

1. I. Schapera, *Tribal Legislation among the Tswana of the Bechuanaland Protectorate* (London, 1943); see generally my *Individual and Community: The Rise of the Polis 800–500 B.C.* (New York, 1986), ch. 2.
2. *A Selection of Greek Historical Inscriptions to the End of the Fifth Century B.C.*, ed. R. Meiggs and D. Lewis (Oxford, 1969), no. 2; the guardians of the law were the *kosmos,* the Demioi, and the Twenty of the *Polis,* though who the latter were is unknown.

3. Thucydides 1. 126; Herodotus 5. 71.

4. Plutarch, *Solon* 8. 1, calls the decree a *nomos;* Diogenes Laertius 1. 46, a *psephisma,* a term later used for legislation by the assembly, but he is too late a source to be trusted.

5. Thucydides 2. 15.

6. See my comments in *The Flawed Mirror* (Lawrence, Ks., 1983), pp. 25ff.

7. R. A. de Laix, *Probouleusis at Athens* (Berkeley, 1973), p. 10; *Constitution of the Athenians* 8. 2; Aristotle, *Politics* 2. 11 1273b 41—1274 a 1.

8. Aeschines 3. 2; E. Ruschenbusch, *Solonos Nomoi* (*Historia, Einzeschrift* 9, 1966), F 101, considers the law doubtful.

9. Against the doubts of Hignett, Will, Sealey, and others see P. J. Rhodes, *The Athenian Boule* (Oxford, 1972), pp. 208-09; de Laix, *Probouleusis,* pp. 13–17.

10. R. W. Wallace, *The Areopagus Council, to 307 B.C.* (Baltimore, 1988), chs. 1–2, denies the whole concept of an aristocratic council before Solon and ascribes its powers to guard the laws to the lawgiver himself (based on Plutarch, *Solon* 19, and *Constitution of the Athenians* 8. 4). Full critique is not warranted in a footnote, but his efforts to discard the clear statements in *Constitution* and even in Aristotle himself (see note 17 below) and to rely to a considerable degree on later sources are scarcely compelling in view of the importance of aristocratic councils (such as the Roman Senate). And why did Solon establish *two* councils?

11. So Georg Busolt, *Griechische Staatskunde,* 2 (3d ed.; Munich, 1926), pp. 842, 850.

12. *Greek Historical Inscriptions,* no. 8; L. H. Jeffery, *Annual of the British School at Athens,* 51 (1956), pp. 157–67, lowered the date.

13. A. Andrewes, *Probouleusis* (Oxford, 1954).

14. Aristotle, *Politics* 2. 12 1274a; *Constitution* 9.

15. As Hansen, *Assembly,* p. 104, notes, Rhodes and other English scholars consider the *heliaea* an arm of the assembly, but both he and many German students dissent. In a paper delivered at the 1989 meetings of the American Philological Association, however, J. Ober advanced evidence that in the fourth century the law courts and assembly were not so strongly contrasted) (e. g., Dinarchus 1. 84; Demosthenes 19. 224, et al.).

16. E. Diehl, *Anthologia Lyrica Graeca* (3d ed.; Leipzig, 1954), fr. 5: *Constitution of the Athenians* 12. 1.

17. Aristotle, *Politics* 2. 12 1273b.

18. Herodotus 1. 59; *Constitution* 14. The tale that he moved a meeting of citizens from the Theseum to the Acropolis so that his aides could seize their arms has little to recommend it (Busolt, *Griechische Staatskunde,* p. 863 n. 2).

19. Herodotus 1. 62.

20. Thucydides 6. 54. There is no need here to survey modern literature on the Pisistratids; see recently F. Frost, "Toward a History of Peisistratid Athens," *The Craft of the Ancient Historian: Essays in Honor of Chester G. Starr,* eds. J. W. Eadie and J. Ober (Lanham, Md., 1985), pp. 57–78.

Chapter II

1. Within the extensive literature on Cleisthenes see especially D. W. Bradeen, "The Trittyes in Cleisthenes' Reforms," *Transactions of the American Philological Association,* 86 (1955), pp. 22–30; D. M. Lewis, "Cleisthenes and Attica," *Historia,* 12 (1963), pp. 22–40; J. Martin, "Von Kleisthenes zu Ephialtes," *Chiron,* 4 (1974), pp. 5–42; C. Meier, "Cleisthenes et le problème politique de la polis grecque," *Revue des droits de l'antiquité,* 3d ser. 20 (1973), pp. 115–59.

2. P. Lévêque et P. Vidal-Naquet, *Clisthène l'Athénien* (Paris, 1964), stress this geographical sensitivity.

3. D. Whitehead, *The Demes of Attica, 508–7—ca. 250 B. C.* (Princeton, 1986), pp. 18–21; J. S. Traill, *Demos and Trittys* (Toronto, 1986).

4. De Laix, *Probouleusis,* pp. 149–53; E. S. Stavely, *Greek and Roman Voting and Elections* (Ithaca, 1972), pp. 38–40.

5. On his efforts to neutralize the influence of cults see Lewis, *Historia,* 12 (1963), pp. 30–35.

6. J. B. Salmon, *Wealthy Corinth* (Oxford, 1984), pp. 207–09, 413–19; N. F. Jones, "The Civic Organization of Corinth," *Transactions of the American Philological Association,* 110 (1980), pp. 161–93; on east Greece, J. M. Cook, *Cambridge Ancient History,* 3. 3 (Cambridge, 1982), pp. 200–01.

7. On possible gerrymandering see G. Daverio Rocchi, "Politica di familia e politica di tribù nella polis ateniese (V secolo)," *Acme,* 24 (1972), pp. 13–44; G. R. Stanton, "The Tribal Reform of Kleisthenes the Alkmeonid," *Chiron,* 14 (1984), pp. 1–41; P. J. Bicknell, "Kleisthenes as Politician," *Historia,* Einzelschrift 19 (1972), pp. 1–53.

8. This thesis is fully developed in C. Meier, *Die Entstehung des*

Politischen bei den Griechen (Frankfort, 1980); see also Meier and P. Veyne, *Kannten die Griechen die Demokratie?* (Berlin, 1988).

9. Herodotus 5. 78.

10. J. D. Lewis, "Isegoria at Athens: When Did It Begin?" *Historia,* 20 (1971), pp. 129–40, dates the right to Solon and Cleisthenes; G. T. Griffith, in *Ancient Society and Institutions* (Oxford, 1966), pp. 115–38, places it about 457/6; A. G. Woodhead, "Isegoria and the Council of 500," *Historia,* 16 (1967), pp. 129–40, essentially accepts Griffith's argument. K. Raaflaub, "Des freien Bürgers Recht der freien Rede," *Studien zur Antiken Sozialgeschichte,* ed. W. Eck et al. (Köln, 1980), pp. 28–34, argues that Cleisthenes at least endorsed *isegoria* for the hoplite class; see also J. Ober, *Mass and Elite in Democratic Athens* (Princeton, 1989), pp. 72–73, 78–79.

11. De Laix, *Probouleusis,* p. 157, is not as certain that the oath contained this clause as are Busolt, *Griechische Staatskunde,* p. 1023, and others.

12. *Constitution of the Athenians* 22. 4; Solon had already legislated against attempted tyranny (*Constitution* 8. 4). See generally M. Ostwald, "The Athenian Legislation against Tyranny and Subversion," *Transactions of the American Philological Association,* 86 (1955), pp. 103–28.

13. Ober, *Mass and Elite,* p. 74; R. Thomsen, *The Origins of Ostracism* (Copenhagen, 1972); E. Vanderpool, "Ostracism at Athens," *Lectures in Honor of Louise Taft Semple* 2 (Cincinnati, 1970), pp. 215–50. A fifteenth-century manuscript asserts that ostracism predates Cleisthenes, but this is certainly debatable; see J. J. Keaney and A. E. Raubitschek, "A Late Byzantine Account of Ostracism," *American Journal of Philology,* 93 (1972), pp. 87–91.

14. J. K. Davies, *Athenian Propertied Families 600–300 B. C.* (Oxford, 1971), p. 375, an invaluable study of the pedigrees as far as known of every Athenian noted in the present work.

15. R. Seager's summation of Meier's *Entstehung* in *Journal of Hellenic Studies,* 102 (1982), pp. 266–67; see also H. W. Pleket, "Isonomia and Cleisthenes, A Note," *Talanta,* 4 (1972), pp. 64–81.

16. Plutarch, *Themistocles* 5. 5; *Aristides* passim; *Cimon* 5. 4, 10. 7; in *Constitution of the Athenians* 23–24 Aristides advises the Athenians to give up the country and live in the city.

17. Plutarch, *Aristides* 13. 1–2, does refer to a plot by some well-to-do Athenians in 479 to subvert democracy in order to put an end to the opposition to the Persians.

18. Herodotus 5. 97.

19. Andocides 1. 43.

20. *Greek Historical Inscriptions,* no. 14; B. D. Meritt, *Hesperia,* 10 (1941), pp. 305–06, restored the reference to the council; de Laix, *Probouleusis,* p. 88, raises the doubt.

21. *Inscriptiones Graecae,* I (3d ed.; Berlin, 1981), no. 4.

22. *Greek Historical Inscriptions,* no. 23, also given in part in Demosthenes 19. 303. C. Habicht, *Hermes,* 88 (1961), pp. 1–35, presents the most powerful attack against its genuineness, but many scholars wish to believe in it, as recently N. G. L. Hammond, *Cambridge Ancient History,* 4 (2d ed.; Cambridge, 1988), pp. 559–63.

23. *Constitution of the Athenians* 22 gives 500; Kenyon in the *editio princeps* argues for 100, the standard number in later days (*Constitution* 8. 1).

24. G. L. Cawkwell, *Journal of Hellenic Studies,* 108 (1988), p. 2, is properly doubtful that the quality of archons declined; see also E. Badian, *Antichthon,* 5 (1971), pp. 319ff.

25. Greek *Historical Inscriptions,* no. 27; this monument was moved by Constantine to his new capital and still stands in Istanbul.

26. Herodotus 7. 161, 140–43.

27. Herodotus 7. 144 uses the technical word *edoxe.*

28. Herodotus 7. 139; Wallace, *Areopagus Council,* p. 78, and many others reject the payment of eight drachmas (*Constitution of the Athenians* 23), but it accords too neatly with the pattern of Athenian coinage to be dismissed.

29. In the abundant literature on these events, C. Hignett, *Xerxes' Invasion of Greece* (Oxford, 1963), and A. J. Podlecki, *The Life of Themistocles* (Montreal, 1975), may suffice.

30. Thucydides 7. 21; cf. 1. 18.

31. D. Whitehead, *The Ideology of the Athenian Metic* (Cambridge, 1971); deliberate refusal by metics to enter public life is illustrated in Euripides, *Suppliants* 888–900.

32. Herodotus 9. 5.

33. J. M. Balcer, "Athenian Politics: The Ten Years after Marathon," *Panathenaia* (Lawrence, Ks., 1979), pp. 27–49.

34. Badian, *Classical Views,* 32 (1988), pp. 304–10, recently has supported Hammond's view (*Historia* 4, [1955], pp. 371–81) that there were two Athenian expeditions in support of the Spartans, but I fail to be convinced that Diodorus' account (11. 63–64) can thus be saved; Badian, p. 316, may be more justified in his assertion that Ephialtes' reforms were not passed in the absence of Cimon, as is usually argued.

35. The standard work on the history of the Athenian empire is
R. Meiggs, *The Athenian Empire* (Oxford, 1972), but one must also
cite the magisterial study of J. de Romilly, *Thucydides and Athenian
Imperialism* (Oxford, 1963).

36. The most recent discussion of Ephialtes' reforms is G. L. Cawkwell, *"Nomophulakia* and the Areopagus," *Journal of Hellenic Studies,*
108 (1988), pp. 1–12, with full references to the limited sources and
modern bibliography including Ruschenbusch's misguided effort to
remove the reforms from Athenian history.

37. Plutarch, *Cimon* 15. 1; Aristotle, *Politics* 5. 4 (the bracketed
interpolation is that of Barker). Cawkwell doubts the renascence of
the authority of the Areopagus but has a good discussion of the elas-
ticity of "guardianship of the laws."

38. Wallace, *Areopagus Council,* pp. 83–87, is rather brief but accepts the removal of *eisangelia* from the jurisdiction of the Areopagus
(on *nomophylakia* see pp. 55–61).

39. Rhodes, *Boule,* c. 4; Ostwald, *Sovereignty,* pp. 50–53; R. Sealey,
"Ephialtes, *Eisangelia,* and the Council," *Classical Contributions:
Studies in Honor of M. F. McGregor* (Locust Valley, 1981), pp. 125–
34. On *euthyna* see J. T. Roberts, *Accountability in Athenian Government* (Madison, 1982).

40. Ostwald, *Sovereignty,* p. 43, sums up the evidence for the development of *dokimasia;* as Wallace, *Areopagus Council,* p. 67, points
out, its origins are of uncertain date.

41. Ostwald, *Sovereignty,* p. 51; M. H. Hansen, *Eisangelia* (Odense,
1975).

42. G. Grote, *History of Greece,* 4 (2d ed.; London, 1869), p. 459;
Busolt, *Griechische Staatskunde,* p. 896; Rhodes, *Boule,* p. 62; on the
dangers see Hansen, *Assembly,* p. 59. Andocides 1. 17 is the first reference to its use; Ober, *Mass and Elite,* p. 95, recently dates it to
427–15.

43. *Constitution of the Athenians* 63–69 fully describes the elaborate procedures in trials. S. Dow, *Harvard Studies in Classical Philology,* 50 (1939), pp. 1–34, displayed masterful ingenuity in his reconstruction of the allotment machines from surviving physical evidence;
J. H. Kroll, *Athenian Bronze Allotment Plates* (Cambridge, Mass.,
1972), has studied the jurors' tickets.

44. Cawkwell, p. 11.

45. Davies, *Athenian Propertied Families,* pp. 455ff.

46. The argument of S. C. Humphreys, *The Family* (London,
1983), pp. 24–25, that the law was designed to check the tendency of

aristocrats to make international marriages may strike one as overly ingenious.

47. See the recent survey by A. Andrewes, "The Opposition to Pericles," *Journal of Hellenic Studies,* 98 (1978), pp. 1–8, though I cannot accept all his views; also H. D. Meyer, *Historia,* 16 (1967), pp. 141–54.

48. Plutarch, *Pericles* 7. 2, 8. 4.

49. de Romilly, *Thucydides and Athenian Imperialism,* p. 79; on pp. 71–73 she adequately discounts the efforts of G. B. Grundy and others to find economic motives for Athenian imperialism.

50. *Greek Historical Inscriptions,* no. 45; my *Athenian Coinage, 480–449 B. C.* (Oxford, 1970), pp. 68–70; the essays by D. M. Lewis and H. B. Mattingly in *Coinage and Administration in the Athenian and Persian Empires,* ed. I. Carradice (BAR International Series no. 343, 1987).

51. *Greek Historical Inscriptions,* no. 40.

52. *Greek Historical Inscriptions,* no. 52; Old Oligarch 1. 17. See also G. E. M. de Ste Croix, "Jurisdiction in the Athenian Empire," *Classical Quarterly,* 11 (1961), pp. 94–112, 268–80.

53. F. Frost, *Journal of Hellenic Studies,* 84 (1964), pp. 69–72, and *Historia,* 13 (1964), pp. 385–99; P. Krentz, *Historia,* 33 (1983), pp. 502–03. E. Derenne, *Les Procès d'impiété intentés aux philosophes à Athènes au Vme et au IVme siècles avant J.-C.* (Liège, 1930), remains useful.

54. Thucydides, 2. 65.

55. A. W. Gomme, *s. v.,* Pericles in *Oxford Classical Dictionary* (2d ed.; Oxford, 1970).

56. One must, however, keep in mind P. A. Stadter's demonstration (*The Speeches in Thucydides* [Chapel Hill, 1973]), that Plutarch in his life of Pericles never cited this speech as reflecting Pericles' outlook even though he refers to it in his moral essays.

57. Plutarch, *Pericles* 7, 9; Thucydides passim; Teleclides frr. 42, 44 (T. Koch, *Comicorum Atticorum Fragmenta* [Leipzig, 1980], p. 220).

58. M. Ostwald, *Autonomia: Its Genesis and Early History* (Chico, Calif., 1982); V. Ehrenberg, *The Greek State* (Oxford, 1960), part 1; my essay, "Athens and Its Empire," *Classical Journal,* 83 (1988), pp. 114–23.

59. Events bearing specifically on the assembly during the Peloponnesian War will appear in later chapters; among the many modern surveys, E. Will, *Le Monde grec et l'Orient: le Ve siècle* (Paris, 1972), pp. 315ff., is a lucid analysis.

Chapter III

1. A. R. W. Harrison, *The Law of Athens: The Family and Property* (Oxford, 1968), pp. 1–60.

2. A. W. Gomme, *The Population of Athens in the Fifth and Fourth Centuries B. C.* (Oxford, 1933), passim, though I have reduced his figure of 115,000 slaves for reasons advanced in the supplementary note to this chapter.

3. A. H. M. Jones, *Athenian Democracy* (Oxford, 1957), pp. 75–76.

4. D. Schaps, *The Economic Rights of Women in Ancient Greece* (Edinburgh, 1979), is a valuable corrective for the Greek world as a whole.

5. *Historical Statistics of the United States* (Bureau of the Census, 1975), pp. 15–16. I am indebted to my colleagues Terrence McDonald and Maris Vinovskis in interpreting these figures.

6. Foreign-born residents are reported in the 1860 census figures, but one cannot be certain how many had received citizenship, which was relatively easy to obtain in the period; in the decade before 1860 there were two and one-half million immigrants, who could vote in Indiana, Michigan, Wisconsin, and Minnesota after their declaration of intent to seek citizenship (D. V. Smith, *Ethnic Voters and the Election of Lincoln,* ed. F. C. Luebke [Lincoln, Neb., n. d.], pp. 1–2).

7. D. V. Smith, p. 13, finds that 77% of Illinois men of voting age cast ballots; if that percentage applied to all states, there were 6,100,000 eligible voters in 1860, a figure not far off from the 6,300,000 suggested in the text above.

8. Paul Kleppner, *Who Voted?* (New York, 1982), p. 5.

9. K. Kourioniotes and H. A. Thompson, 'The Pnyx in Athens," *Hesperia,* 1 (1932), pp. 90–217; Thompson, "The Pnyx in Models," *Hesperia,* Supp. 19 (1982), pp. 133–47.

10. Jones, *Athenian Democracy,* p. 8, suggests the proportions 1:2 for hoplites and *thetes.*

11. Thucydides 8. 72.

12. Hansen, *Athenian Assembly,* p. 9.

13. As cited by Neil Spitzer, *Atlantic Monthly,* November 1988, p. 20.

14. The number of 40 meetings a year has been questioned at least for the fourth century; cf. E. M. Harris, "How Often Did the Athenian Assembly Meet?" *Classical Quarterly,* 36 (1986), 363–77; opposed by Hansen, *Assembly,* pp. 196–97, and *Greek Roman and Byzantine*

Studies, 28 (1987), pp. 51–58; F. Mitchel and Hansen, "The Number of *Ecclesiai* in Fourth-century Athens," *Symbolae Osloenses,* 59 (1984), pp. 13–19.

15. Aristotle, *Politics* 6. 4 1318b, also notes the geographical problem of rural participation in government.

16. Gomme, *Population of Athens,* pp. 37–38; see also the appendix in Jones, *Athenian Democracy,* pp. 161–80.

17. E. g., Hansen, *Assembly,* p. 11.

18. Xenophon, *Memorabilia* 3. 7. 6; Aristotle, *Politics* 4. 6 1293a, 4. 12 1296b, 6. 4 1319b.

19. Plato, *Republic* 8. 565a. The slant of E. Kluwe, "Die soziale Zusammensetzung der athenische Ekklesia und ihr Einfluss auf politische Entscheidungen," *Klio,* 58 (1976), pp. 295–323 and 59 (1977), pp. 45–81, may be gathered from its title, but it is a generally sound survey. The question of popular role in Athenian public life has recently been treated from different points of view by R. K. Sinclair, *Democracy and Participation in Athens* (Cambridge, 1988), and L. B. Carter, *The Quiet Athenian* (Oxford, 1986), both largely after our period.

20. Jones, *Athenian Democracy,* p. 109.

21. M. I. Finley, *Ancient Slavery and Modern Ideology* (Penguin, 1980), p. 80, suggests the figure of 60,000 slaves; G. E. M. de Ste Croix, *Classical Review,* 7 (1957), pp. 54ff., estimates 60,000–80,000, as does S. Lauffer, *Die Bergwerkssklaven von Laureion* (Mainz, 1955–56), 2, pp. 190–96.

22. Thucydides 7. 27; Finley, *The Ancient Economy* (Berkeley, 1973), p. 24, calls this a guess; Sealey, *Athenian Republic,* p. 9, thinks the Spartans did engage in deliberate numeration.

23. Jones, *Athenian Democracy,* pp. 76–79.

24. Eleusis, *Inscriptiones Graecae,* II (2d ed.), no. 1672; Erechtheum, *Inscriptiones Graecae,* I (3d ed.), no. 476.

25. K. M. Stampp, *The Peculiar Institution* (New York, 1956), pp. 29–30, cited by Finley, *Economy and Society in Ancient Greece* (Penguin, 1983), p. 102; Ferguson, *Greek Imperialism,* p. 61.

26. G. E. M. de Ste Croix, *Ancient Society and Institutions,* pp. 109–14.

27. *Politics* 6. 8 1323a. E. Wood, "Agricultural Slavery in Classical Athens," *American Journal of Ancient History,"* 8 (1983), pp. 1–47, is a safer guide than M. H. Jameson, "Agriculture and Slavery in Classical Athens," *Classical Journal,* 73 (1977), pp. 122–41.

28. As far as I know my essay, "An Overdose of Slavery," *Journal*

of Economic History, 15 (1958), pp. 17–32 (reprinted in my *Essays on Ancient History,* eds. A. Ferrill and T. Kelly [Leiden, 1979], pp. 43–58), has been favorably noticed only by Badian, *Craft of the Ancient Historian,* p. 13; J. A. Lencman, *Die Sklaverei in mykenischen und Homerischen Griechenland* (Wiesbaden, 1966), p. 93, found me one of "die verbissenen Gegner des Marxismus," a charge to which I would reply that I was not thinking of Marxism at all in describing the actual character and spread of Greek slavery. The historian must put valid evidence before theory of any source.

Chapter IV

1. Aristotle, *Politics* 6. 8 1322b, 7. 8 1328b; G. Rudé, *Europe in the Eighteenth Century* (London, 1972), p. 103, also specifies as one of the main duties of the *Ancien Régime* the protection of "the established church."

2. W. K. Pritchett, *Hesperia,* 25 (1956), pp. 215–17, lists a stool (*diphros*) apparently for one obol, though he suggests it might actually be one drachma; such sporadic precision occurs in other accounts, as in the amount spent at Eleusis for repairing the sandals of a slave (*Inscriptiones Graecae,* II [2d ed.], no. 1673, lines 45–51). On the investigation of the profanation see D. MacDowell, *Andokides on the Mysteries* (Oxford, 1962); D. M. Lewis, "After the Profanation of the Mysteries," *Ancient Society and Institutions,* pp. 177–91.

3. See my essay, "Religion and Patriotism in Fifth-century Athens," *Panathenaia,* pp. 11–25.

4. See generally H. W. Parke, *Festivals of the Athenians* (Ithaca, 1977), pp. 17, 42; Ostwald, *Sovereignty,* pp. 139–40.

5. *Greek Historical Inscriptions,* no. 44.

6. *Greek Historical Inscriptions,* no. 73.

7. Cf. the questions put to Glaucon by Socrates as reported in Xenophon, *Memorabilia* 3. 6; see generally my essay, "Greek Administration," in *Civilization of the Ancient Mediterranean,* ed. M. Grant and R. Kitzinger (New York, 1988), pp. 631–47.

8. A. Andreades, *A History of Greek Public Finance,* 1 (Cambridge, Mass., 1933), p. 243.

9. The former survives only on papyrus; see C. W. Fornara, *Archaic Times to the End of the Peloponnesian War* (Baltimore, 1977), no. 94 A-B, and B. D. Merritt, H. T. Wade-Gery, *The Athenian Tribute*

Lists, 4 vols. (Cambridge, Mass., 1939–53), at vol. 2, p. 61. The currency decree is known in several copies; see *Greek Historical Inscriptions,* no. 45, and the full bibliography in my *Athenian Coinage,* pp. 66–68.

10. *Greek Historical Inscriptions,* no. 69, at pp. 196–97.

11. Aristotle, *Politics,* 3. 6 1279a; Herodotus 8. 4. In the fourth century Hypereides 5. 25 asserted that Demosthenes and Demades gained more than 60 talents from securing the passage of decrees.

12. R. S. Stroud, "An Athenian Law on Silver Coinage," *Hesperia,* 43 (1974), pp. 157–88. For other examples see W. Dittenberger, *Sylloge Inscriptionum Graecarum* (3d ed.; Leipzig, 1915–24), no. 218 (Olbia); M. N. Tod, *A Selection of Greek Historical Inscriptions,* 2 (Oxford, 1948), no. 112 (Mytilene and Phocaea).

13. So the Old Oligarch 3. 2 observed that "the council supervised the dockyards (*neoria*)."

14. B. Jordan, The *Athenian Navy in the Classical Period* (Berkeley, 1975), pp. 61–63.

15. Rhodes, *Boule,* pp. 43–46.

16. Thucydides 2. 22; Plutarch, *Pericles* 33. 5–6; as Hansen, *Assembly,* p. 22, points out, Pericles may have acted constitutionally, for more than 70 days could elapse between major meetings of the assembly; but the longest Spartan invasion lasted 40 days. That in 425 was the shortest, only 15 days (Thucydides 4. 6).

17. Plutarch, *Nicius* 7. 5; Aristophanes, *Knights* 164–67.

18. Thucydides 1. 45, 49. 7.

19. Thucydides 6. 8, 6. 26.

20. Demosthenes 4. 26.

21. Andocides 1. 83.

22. Thucydides 8. 45.

23. Ferguson, *Greek Imperialism,* p. 51.

24. Ferguson argued that the council could not act without first referring a matter to the assembly, but Rhodes, *Boule,* p. 166, cites examples where the council began an accusation.

25. *Constitution of the Athenians* 43. 5.

26. Plutarch, *Pericles* 31.

27. Diodorus 13. 6; Ostwald, *Sovereignty,* p. 276.

28. Rhodes, *Boule,* pp. 162–63, notes that there was not a finite list of offenses for this charge; see the full discussion in Busolt, *Griechische Staatskunde,* 2, p. 848, note 3.

29. Ostwald, *Sovereignty,* pp. 28ff.; C. Hignett, *A History of the*

Athenian Constitution to the End of the Fifth Century B. C. (Oxford, 1952), p. 154 (Hignett's views are often too idiosyncratic to need frequent citation here; Ostwald, Rhodes, and others are safer guides).

30. Herodotus 6. 132–36; Cornelius Nepos, *Miltiades* 8, on which see recently A. C. Dionisotti, *Journal of Roman History,* 78 (1988), pp. 47–48.

31. Xenophon, *Hellenica* 1. 1.

32. It is not always remembered that at Marathon Athenian slaves stood beside the Athenian and Plataean hoplites and died in the battle (Pausanias 1. 32. 3, who saw the grave of the slaves). The trust which Athenians could thus place in their slaves must help to offset the usual picture of the brutality of ancient slavery (based largely on Roman times), though one must keep in mind the ruthless exploitation of slaves in the mines at Laurium. Even so, at Athens according to the Old Oligarch 1, "a blow is illegal, and a slave will not step aside to let you pass him in the street."

33. Xenophon, *Hellenica* 1. 7. 12; Ostwald, *Sovereignty,* pp. 434ff.

Chapter V

1. G. T. Griffith gives a magnificent assessment of Philip's career in N. G. L. Hammond and Griffith, *A History of Macedonia,* 2 (550–336 B. C.) (Oxford, 1979).

2. Demosthenes 18. 169–70.

3. One of the few blemishes in H. D. F. Kitto, *The Greeks* (Harmondsworth, 1951), p. 33, is his far too optimistic picture of Greek longevity. Arthur Schlesinger, Sr., *The Birth of the Nation* (New York, 1968), p. 200, also cited a series of long lives in the early Republic (Jefferson, 83; John Adams, 91; etc.) but pointed out that life expectancy in Massachusetts and New Hampshire according to church records was only 35 years; of Cotton Mather's 16 children but one survived him. See generally my *Economic and Social Growth,* pp. 40–43.

4. Jones, *Athenian Democracy,* p. 107.

5. Aeschines, *Against Timarchus* 35 (quoted by Rhodes, *Boule,* p. 141).

6. Plato, *Protagoras* 319b. Aristotle, *Politics* 3. 11 1281b put it that while each individual might be of poor quality the sum may exceed that of a select few—a basic point in the justification of democracy in any society.

7. K. J. Dover, "Anapsephesis in Fifth-century Athens," *Journal of Hellenic Studies,* 75 (1955), pp. 17–20.

8. Ober, *Mass and Elite*, p. 79, who notes that most citizens never exercised their right but that nonetheless it existed *in posse* (pp. 296–98).

9. As in the Mitylenean debate (Thucydides 3. 49), where the show of hands was almost even.

10. Hansen, *Assembly*, p. 43.

11. Aristotle, *Politics* 2. 9 1271a.

12. A. S. Henry, *The Prescripts of Athenian Decrees, Mnemosyne*, Supp. 49 (1977), observes (pp. 104–05) the free, almost erratic manner in which decrees were recorded and then carved on stone.

13. Thucydides 2. 65.

14. Plutarch, *Nicias* 12. 1 and *Alcibiades* 17. 3; as I have noted in *Political Intelligence in Classical Greece, Mnemosyne*, Supp. 31 (1974), few expeditions were launched in ancient Greece with greater forethought and care in preparation, which did not prevent disaster.

15. Ostwald, *Sovereignty*, pp. 226–29.

16. Thucydides 8. 47. Ostwald, *Sovereignty*, pp. 346ff., discusses the developments in 411 in detail; Wallace, *Areopagus Council*, pp. 131–44, defined its place in the debates of 411 and 404 and thereafter in Isocrates, *Areopagiticus*, and later works, which fall outside the confines of the present study.

17. Thucydides 8. 48.

18. Thucydides 8. 53.

19. Thucydides 8. 68.

20. Thucydides 8. 97.

21. Plutarch, *Alcibiades* 23; cf. Thucydides' judgment 6. 15 and the recent remarkable analysis by S. Forde, *The Ambition to Rule: Alcibiades and the Politics of Imperialism in Thucydides* (Ithaca, 1989).

22. Xenophon, *Hellenica* 2. 3; P. Krentz, *The Thirty at Athens* (Ithaca, 1982), is a rounded and judicious study.

23. A. Fuks, *The Ancestral Constitution* (London, 1953); M. I. Finley, *The Ancestral Constitution*, his inaugural lecture at Cambridge in 1971.

24. Andocides, *On the Mysteries* 83–84, quotes the decree at length.

25. Xenophon, *Hellenica* 2. 4. 43.

26. Ostwald, *Sovereignty*, p. 497.

27. Andocides, *On the Mysteries* 87; cf. Demosthenes 23. 87; Ostwald, *Sovereignty*, passim.

28. *Constitution of the Athenians* 41; cf. Aristotle, *Politics* 4. 4 1292a. Hansen, *Assembly*, pp. 96–97, judges that the *demos* is here considered as meaning the common people; thereafter Hansen notes

various limitations on the power of the assembly in the later fourth century.

29. As suggested too briefly by Jones, *Athenian Democracy,* p. 41; so too Ostwald, *Sovereignty,* p. 343.

30. Thucydides 6. 39; Herodotus 3. 80.

31. Aristotle, *Politics* 2. 12 1273b.

32. Hansen, *Ecclesia,* pp. 137, 211ff.

33. Forde, *Ambition to Rule,* p. 74, "The greatest Athenian statesmen all seem to acknowledge the need for at least some deception in dealing with the *demos.*" On tension see pp. 7–9, 29–32, 94–95, 147–51; and in the fourth century Ober, *Mass and Elite,* passim.

34. W. R. Connor, *The New Politicians of Fifth-century Athens* (Princeton, 1971).

35. R. J. Hopper, *The Basis of Athenian Democracy* (Sheffield, 1957), quoted by Whitehead, *The Demes of Attica,* p. xvii.

36. Whitehead, pp. 317–18.

37. Whitehead, pp. 237–40.

38. Whitehead, p. 319; R. Osborne, *Demos: The Discovery of Classical Attika* (Cambridge, 1985), also finds a distinction between deme and *polis* but takes the deme to be the basic building block for Athenian democracy (pp. 91–92).

39. Whitehead, pp. 309–11; Connor, *New Politicians,* pp. 119–28; J. K. Davies, *Wealth and the Power of Wealth in Classical Athens* (New York, 1981).

40. S. Dow, "Companionable Associates in the Athenian Government," *In Memoriam Otto Brendel* (Mainz, 1976), pp. 69–84, finds some groups of cronies in the appointment of assessors (*paredroi*) by the archons.

41. As suggested by G. R. Stanton and P. J. Bicknell, "Voting in Tribal Groups in the Athenian Assembly," *Greek Roman and Byzantine Studies,* 28 (1987), pp. 51–92; Ober, *Mass and Elite,* pp. 132–33, is doubtful and cites Theophrastus, *Characters* 26, on the "socially undifferentiated mass" at least in the fourth century. See also Hansen, *Assembly,* pp. 40–41, though A. Boegehold, *Hesperia,* 32 (1963), pp. 373–74, notes the report of Kourouniotes and Thompson of some evidence in the fifth century for physical divisions of the Pnyx.

42. Plutarch, *Pericles* 11. 2–3; Aristophanes, *Ecclesiazusae* 87, 98.

43. Thucydides 8. 54, 65, 66; W. G. Forrest, "An Athenian Generation Gap," *Yale Classical Studies,* 24 (1975), 37–52, is the most recent discussion of the people involved in the upheaval of 411. A friendly critic points out that youths at symposia were given to singing

songs in praise of the tyrannicides, but how far this was an expression of political stance in the fifth century seems unclear.

44. Jones, *Athenian Democracy*, p. 131; see, for example, *Ecclesiazusae* 197–98.

45. Grote, *History of Greece*, 3, p. 354; de Laix, *Probouleusis*, p. viii, notes other views; Ehrenberg, *Greek State*, p. 63, called the council "the instrument through which the Demos ruled."

46. E. Posner, *Archives in the Ancient World* (Cambridge, Mass., 1972).

47. Thucydides 4. 118, 5. 45, 5. 47; in Aristophanes, *Thesmophoriazusae* 372ff, 943, the council has a significant role.

48. Rhodes, *Boule*, pp. 52–81.

49. De Laix, *Probouleusis*, pp. 107–08, 192–93; in his review Rhodes, *Journal of Hellenic Studies*, 94 (1974), pp. 232–33, feels that de Laix overstresses the role of the council at times.

50. Hansen, *Assembly*, p. 37; cf. Jones, *Athenian Democracy*, pp. 105, 118.

51. De Laix, *Probouleusis*, pp. 174–76; Jones, *Athenian Democracy*, pp. 90–92, correctly points out that at Athens there was never any effort to redistribute lands or to cancel debts (after the reforms of Solon).

52. E. g., Forde, *Ambition to Rule*, p. 102, "It is an undeniable defect of democracy as a regime that it is rather too prone to the influence of leaders who turn it, moblike, to the bad." To counter this sweeping assertion as regards modern times would take us far afield.

53. Thucydides 2. 37–40.

Bibliography

Quotations of ancient authors are usually, but not always drawn from Penguin or Loeb translations; I have sometimes modified them to reflect more precisely the Greek text. On two topics more specific comment is necessary:

a) the papyrus containing the *Athenaion Politeia* was found in Upper Egypt and sold by an American missionary to the British Museum. Its first edition was by F. Kenyon; I have used that by J. E. Sandys (London, 1912), which has excellent notes. F. von Fritz and E. Kapp published a good translation (New York, 1950); the most recent commentary is by P. J. Rhodes (Oxford, 1981).

b) Inscriptions I have cited wherever possible from R. Meiggs and D. Lewis, *A Selection of Greek Historical Inscriptions to the End of the Fifth Century B. C.* (Oxford, 1969), but at times have had to turn to *Inscriptiones Graecae*, especially vol. I (3d ed.; Berlin, 1981). Many have been translated in C. W. Fornara, *Archaic Times to the End of the Peloponnesian War* (Baltimore, 1977).

Collected Essays

Ancient Society and Institutions: Studies Presented to Victor Ehren-berg (Oxford, 1966).
Civilization of the Ancient Mediterranean, ed. M. Grant and R. Kitz-inger (New York, 1988).

The Craft of the Ancient Historian: Essays in Honor of Chester G. Starr, ed. J. W. Eadie and J. Ober (Lanham, Md., 1985).
Panathenaia, ed. T. E. Gregory and A. J. Podlecki (Lawrence, Ks., 1979).

Other Modern Works (N. B. Articles and general works on Athenian democracy are omitted here.)

Andrewes, A., *Probouleusis* (Oxford, 1954).
Bleicken, J., *Die athenische Demokratie* (Paderborn, 1985).
Busolt, G., *Griechische Staatskunde,* 2 (3d ed.; Munich, 1926).
Carter, L. B., *The Quiet Athenian* (Oxford, 1986).
Connor, W. R., *The New Politicians of Fifth-century Athens* (Princeton, 1971).
Davies, J. K., *Athenian Propertied Families 600–300 B. C.* (Oxford, 1971).
Davies, J. K., *Wealth and the Power of Wealth in Classical Athens* (New York, 1981).
Ehrenberg, V., *The Greek State* (Oxford, 1960).
Ferguson, W. S., *Greek Imperialism* (Boston, 1913).
Ferguson, W. S., *The Treasurers of Athena* (Cambridge, Mass., 1932).
Finley, M. I., *The Ancestral Constitution* (Cambridge, 1971).
Forde, S., *The Ambition to Rule: Alcibiades and the Politics of Imperialism in Thucydides* (Ithaca, 1989).
Fornara, C. W., *The Athenian Board of Generals from 501 to 404, Historia,* Einzelschrift 16 (1971).
Fuks, A., *The Ancestral Constitution* (London, 1953).
Grote, G., *History of Greece* (2d ed.; London, 1869).
Hansen, M. H., *The Athenian Assembly* (Oxford, 1987).
Hansen, M. H., *The Athenian Ecclesia* (Copenhagen, 1983).
Hansen, M. H., *Eisangelia* (Odense, 1975).
Headlam, J. W., *Election by Lot at Athens* (2d ed. with notes by D. C. MacGregor; Cambridge, 1933).
Henry, A. S., *The Prescripts of Athenian Decrees, Mnemosyne,* Supp. 49 (1977).
Hignett, C., *A History of the Athenian Constitution to the End of the Fifth Century B. C.* (Oxford, 1952).
Jones, A. H. M., *Athenian Democracy* (Oxford, 1957).
Jordan, B., *The Athenian Navy in the Classical Period* (Berkeley, 1975).
Krentz, P., *The Thirty at Athens* (Ithaca, 1982).

de Laix, R. A., *Probouleusis at Athens* (Berkeley, 1973).

Meier, C., *Die Entstehung des Politischen bei den Griechen* (Frankfurt, 1980).

Meiggs, R., *The Athenian Empire* (Oxford, 1972).

Ober, Josiah, *Mass and Elite in Democratic Athens: Rhetoric, Ideology, and the Power of the People* (Princeton, 1989).

Osborne, R., *Demos: The Discovery of Classical Attika* (Cambridge, 1985).

Ostwald, M., *Autonomia: Its Genesis and Early History* (Chico, Calif., 1982).

Ostwald, M., *From Popular Sovereignty to the Sovereignty of Law: Law, Society, and Politics in Fifth-century Athens* (Los Angeles, 1987).

Rhodes, P. J., *The Athenian Boule* (Oxford, 1972).

Roberts, J. T., *Accountability in Athenian Government* (Madison, 1982).

de Romilly, J., *Thucydides and Athenian Imperialism* (Oxford, 1963).

Sealey, R., *The Athenian Republic: Democracy or the Rule of Law?* (University Park, 1987).

Sinclair, R. K., *Democracy and Participation in Athens* (Cambridge, 1988).

Starr, C. G., *Athenian Coinage, 480–449 B. C.* (Oxford, 1970).

Starr, C. G., *The Economic and Social Growth of Early Greece, 800–500 B. C.* (New York, 1977).

Starr, C. G., *Individual and Community: The Rise of the Polis, 800–500 B. C.* (New York, 1986).

Staveley, E. S., *Greek and Roman Voting and Elections* (London, 1972).

Thomsen, R., *The Origins of Ostracism* (Copenhagen, 1972).

Traill, J. S., *Demos and Trittys* (Toronto, 1986).

Wallace, R. W., *The Areopagus Council, to 307 B. C.* (Baltimore, 1989).

Whitehead, D., *The Demes of Attica, 508/7–ca. 250 B. C.* (Princeton, 1986).

Index